T0291946

NETWORK CODING

To access full color downloadable figures that feature in the book please
visit www.elsevierdirect.com/companions/9780123809186

NETWORK CODING
Fundamentals and Applications

Edited by

MURIEL MÉDARD
Massachusetts Institute of Technology, Cambridge, MA, USA

ALEX SPRINTSON
Texas A&M University, College Station, TX, USA

AMSTERDAM • BOSTON • HEIDELBERG • LONDON
NEW YORK • OXFORD • PARIS • SAN DIEGO
SAN FRANCISCO • SINGAPORE • SYDNEY • TOKYO

Academic Press is an imprint of Elsevier

Academic Press is an imprint of Elsevier
The Boulevard, Langford Lane, Kidlington, Oxford, OX5 1GB
225 Wyman Street, Waltham, MA 02451, USA

First published 2012

Copyright © 2012 Elsevier Inc. All rights reserved.

No part of this publication may be reproduced or transmitted in any form or by any means, electronic
or mechanical, including photocopying, recording, or any information storage and retrieval system,
without permission in writing from the publisher. Details on how to seek permission, further
information about the Publisher's permissions policies and our arrangement with organizations such as
the Copyright Clearance Center and the Copyright Licensing Agency, can be found at our website:
www.elsevier.com/permissions.

This book and the individual contributions contained in it are protected under copyright by the
Publisher (other than as may be noted herein).

Notices
Knowledge and best practice in this field are constantly changing. As new research and experience
broaden our understanding, changes in research methods, professional practices, or medical treatment
may become necessary.

Practitioners and researchers must always rely on their own experience and knowledge in evaluating
and using any information, methods, compounds, or experiments described herein. In using such
information or methods they should be mindful of their own safety and the safety of others, including
parties for whom they have a professional responsibility.

To the fullest extent of the law, neither the Publisher nor the authors, contributors, or editors, assume
any liability for any injury and/or damage to persons or property as a matter of products liability,
negligence or otherwise, or from any use or operation of any methods, products, instructions, or ideas
contained in the material herein.

British Library Cataloguing in Publication Data
A catalogue record for this book is available from the British Library.

Library of Congress Number: 2011933932

ISBN: 978-0-323-16415-3

For information on all Academic Press publications
visit our website at *www.elsevierdirect.com*

Printed and bound in the United States
12 13 14 15   10 9 8 7 6 5 4 3 2 1

Working together to grow
libraries in developing countries

www.elsevier.com | www.bookaid.org | www.sabre.org

ELSEVIER   BOOK AID
International   Sabre Foundation

CONTENTS

10.   **Network Coding in Disruption Tolerant Networks**   **267**

Xiaolan Zhang, Giovanni Neglia, and Jim Kurose

The network coding area was initiated at the turn of the millennium in a seminal paper by Ahlswede, Cai, and Yeung, and since then has captivated the researchers and practitioners around the world. Recent research efforts shed light on many exciting network coding problems, fundamental capacity limits, and performance gains. The researchers were able to design capacity-achieving network codes for several important practical settings, establish computational complexity of several fundamental network coding problems, and devise efficient network coding algorithms. In addition, the researchers discovered deep connections between network coding and other areas of networking, complexity theory, graph theory, matroid theory, coding theory, and information theory.

Despite significant progress on the theoretical front, practical applications of network coding are only beginning to emerge. While the network coding technique has been successfully applied in many areas, most notably content distribution systems, there is a significant untapped potential for leveraging network coding for improving performance in wireless networks, storage networks, disruption-tolerant networks, and other areas. In addition, the network coding technique has a significant potential to improve security, robustness, manageability, and support of Quality of Service in wired and wireless environments.

This book provides an accessible tutorial introduction and a comprehensive survey of practical applications of network coding in various areas of networking and distributed computing. Our goal is to expose the richness of network coding applications and present new exciting opportunities in network coding research. The book consists of ten chapters written by recognized experts in the field. Each chapter focuses on a specific application area and is written in a clear way, using minimum mathematical notation, so every reader familiar with the basic networking concepts will be able to understand the material. Each chapter provides a comprehensive literature survey with references to the most relevant work in the area for future reading.

Chapter 1 is a tutorial-style introduction to the basics of network coding and provides an overview of the network information flow, linear network coding, the polynomial-time algorithm for code construction, random network coding, coding advantage, subspace transmissions, and nonlinear network codes.

Chapter 2 focuses on the key principles that form the foundation of the network coding implementation in wireless networks. The chapter discusses the principles of opportunistic listening and coding, as well as the details of the forwarding architecture and network coding performance gains. The chapter also describes the principles of analog network coding.

Chapter 3 discusses the applications of network coding in peer-to-peer networks with the focus on content distribution and multimedia streaming. The chapter explores the possible design space of peer-to-peer systems with network coding, and shows the advantages of using the coding technique in this setting.

Chapter 4 discusses the implementation of network coding on commercial mobile platforms. The chapter describes basic system components and design choices as well as practical problems and performance issues associated with implementation of network coding algorithms in mobile phones.

Chapter 5 discusses the applications of network coding in LTE networks. The authors focus on user-cooperation schemes for erasure recovery, and show that leveraging network coding techniques yields significant performance gains.

Chapter 6 focuses on mobile *ad hoc* networks and describes a fully implemented communication system based on a network coding protocol stack. The chapter examines practical implementation issues associated with implementation of network coding on an embedded system and presents results from field experiments.

Chapter 7 focuses on applications of network coding for improving network security. The chapter covers passive adversaries who only have eavesdropping capabilities and active adversaries who can perform both eavesdropping and jamming. The chapter also discusses adversaries with limited computational power.

Chapter 8 discusses connections between network coding, source coding, and data compression.

Chapter 9 focuses on the performance of network coding in wireless networks under several scaling regimes, including the network size, the coding window size, the number of flows in the network, and the application delay constraints. The chapter derives and analyzes scaling laws that help quantify the performance of network coding schemes in large networks.

Chapter 10 focuses on the application of network coding in delay and disruption tolerant networks. The chapter discusses unique challenges that arise in the design of such networks and shows that the network coding technique is instrumental in addressing these challenges. The chapter covers transmission and buffer management, and recovery schemes, as well as coding benefits in energy efficiency and improving delay-transmission tradeoffs.

TARGET AUDIENCE

The book is intended for researchers and practitioners who are interested in the general area of network coding and its applications in various areas of communication networks. We assume that the readers of this book have only a general background in networking, with no prior exposure to the network coding techniques or applications of network coding. This book will be ideal for a networking professional who would like to gain basic knowledge in this area and learn about the application of this technique. The book will also be valuable for graduate students who would like to conduct research on network coding. The book can be used as a textbook or a reference book in courses that focus on network coding or advanced coding technologies.

Muriel Médard
Massachusetts Institute of Technology, Cambridge, MA, USA

Alex Sprintson
Texas A&M University, College Station, TX, USA

ACKNOWLEDGMENTS

We are grateful to all contributors for their dedication, enthusiasm, and commitment in writing and revising the chapters. Clearly, this book would not be possible without their very generous efforts. We are also thankful to Danilo Silva for his help in reviewing the chapters; Tim Pitts, Naomi Robertson, Charlotte Kent and Lisa Jones from Academic Press for their support, encouragement, flexibility, understanding, and help with all aspects of the editorial work. Last but not least we are thankful to our spouses and families for their encouragement, sacrifice in giving us free time to work on the project, and providing inspiration.

MURIEL MÉDARD

Muriel Médard received B.S. degrees in Electrical Engineering and Computer Science (EECS), in Mathematics and in Humanities, as well as M.S. and Sc.D. degrees EE, all from the Massachusetts Institute of Technology (MIT).

She is a Professor in the EECS Department at MIT. She was previously an Assistant Professor at the University of Illinois Urbana-Champaign and a Staff Member at MIT Lincoln Laboratory. Her research interests are in network coding, as well as optical and wireless networks.

Muriel has served as an Associate Editor (AE) for the Optical Communications and Networking Series of the *IEEE Journal on Selected Areas in Communications* (JSAC), for the *IEEE Transactions on Information Theory* (IT), and for the OSA *Journal of Optical Networking*. She has been Guest Editor for the *IEEE/OSA Journal of Lightwave Technology* (JLT), for IT (twice), for JSAC, and for the *IEEE Transactions on Information Forensics and Security*. She is an AE for JLT. She has served as Technical Program Co-chair of IEEE ISIT, of WiOpt, and of ACM CoNEXT.

She is a recipient of the 2009 IEEE William R. Bennett Prize in the Field of Communications Networking, the 2009 IEEE Joint Information Theory/Communications Society Paper Prize, the 2002 IEEE Leon K. Kirchmayer Prize Paper Award, and the Best Paper Award at the Fourth International Workshop on the Design of Reliable Communication Networks, 2003. She received an NSF CAREER Award in 2001 and the 2004 Harold E. Edgerton Faculty Achievement Award at MIT. She was named a 2007 Gilbreth Lecturer by the National Academy of Engineering. She has served on the Board of Governors of the IEEE Information Theory Society for several years, currently as First Vice-President.

ALEX SPRINTSON

Alex Sprintson received the B.Sc. degree (summa cum laude), M.Sc., and Ph.D. degrees in electrical engineering from the Technion–Israel Institute of Technology, Haifa, in 1995, 2001, and 2003, respectively.

Alex is an Associate Professor with the Department of Electrical and Computer Engineering, Texas A&M University, College Station. From 2003 to 2005, he was a Postdoctoral Research Fellow with the California Institute of Technology, Pasadena. His research interests lie in the general area of communication networks with a focus on network coding, network survivability and robustness network algorithms, and QoS routing.

Dr. Sprintson received the Viterbi Postdoctoral Fellowship and the NSF CAREER award. He is an Associate Editor of the *IEEE Communications Letters and Computer Networks Journal*. He has been a member of the Technical Program Committee for IEEE Infocom 2006–2012.

LIST OF CONTRIBUTORS

Frank R. Kschischang   **1**
Department of Electrical and Computer Engineering, University of Toronto, Toronto, Canada

Dina Katabi   **39**
Computer Science & Artificial Intelligence Lab, MIT, MA, USA

Sachin Katti   **39**
Department of Electrical Engineering, Stanford University, CA, USA

Hariharan Rahul   **39**
Computer Science & Artificial Intelligence Lab, MIT, MA, USA

Chen Feng   **61**
Department of Electrical and Computer Engineering, University of Toronto, Toronto, Canada

Baochun Li   **61**
Department of Electrical and Computer Engineering, University of Toronto, Toronto, Canada

Janus Heide   **87, 115**
Department of Electronic Systems, Aalborg University, Aalborg, Denmark

Morten V. Pedersen   **87, 115**
Department of Electronic Systems, Aalborg University, Aalborg, Denmark

Frank H.P. Fitzek   **87, 115**
Department of Electronic Systems, Aalborg University, Aalborg, Denmark

Torben Larsen   **87**
Department of Electronic Systems, Aalborg University, Aalborg, Denmark

Qi Zhang   **115**
Aarhus University, Aarhus, Denmark

Jorma Lilleberg   **115**
Renesas Mobile Corporation, Oulu, Finland

Kari Rikkinen   **115**
Renesas Mobile Corporation, Oulu, Finland

Victor Firoiu   **141**
Advanced Information Technologies, BAE Systems, Burlington, MA, USA

Greg Lauer   **141**
BBN Technologies, Cambridge, MA, USA

LIST OF FIGURES

LIST OF TABLES

An Introduction to Network Coding

Frank R. Kschischang

Department of Electrical and Computer Engineering, University of Toronto, Toronto, Canada

Contents

Abstract

A tutorial introduction is given to network coding, with particular emphasis on the single-source multicasting problem. Many now-standard results are reviewed; in particular: (a) the multicast capacity of a network is given by the least-capacity cut separating the source from any of its destinations; (b) the multicast capacity is achievable with linear network coding over a finite field of sufficiently large size, and

Network Coding. DOI: 10.1016/B978-0-12-380918-6.00001-9
Copyright © 2012 Elsevier Inc. All rights reserved.

there exists a computationally efficient algorithm to find a set of capacity-achieving linear network coding coefficients; (c) the multicast capacity is achieved with high probability over a sufficiently large field when the linear network coding coefficients are chosen at random. Also discussed is the practically important case of noncoherent random network coding, where neither the source nor any of the destinations has *a priori* knowledge of the specific linear network coding operations performed by the network. Finally, several "diabolical" counterexamples are discussed that demonstrate that linear network coding is not sufficient to achieve maximum throughput for general network flow problems.

Keywords: Graphs and networks, multicast capacity, linear network coding, linear information flow algorithm, random linear network coding, noncoherent network coding, nonlinear network coding.

1. THE BUTTERFLY NETWORKS

Network coding is based on a simple, yet far-reaching, idea: in a packet network, rather than simply routing packets, intermediate nodes may compute and transmit *functions* of the packets that they receive. The following examples, the first of which is given in the seminal paper [1] of Ahlswede, Cai, Li and Yeung (in which the field of network coding originated), illustrate some of the advantages that such intermediate coding may provide.

The directed graph shown in Fig. 1.1(a)—the so-called "butterfly network"—models a packet network. In this network, a single source, s, wishes to multicast information to two destinations, t and u. Each directed edge represents an error-free packet channel, capable of delivering, in each channel use, a single packet of m bits. The source wishes to communicate with the two destinations at as high a rate as possible. A key feature of this network is that it contains a bottleneck link: the edge (b,c) directed from node b to node c in Fig. 1.1(a).

A "routing solution" to this problem is shown in Fig. 1.1(b). In time slot 1, the source emits two packets, x and y, routing packet x to both destinations t and u, and routing packet y only to u. In time slot 2, the source emits packets y and z, this time routing packet z to both destinations, and routing packet y only to t. At the end of the two time slots, both destinations have received all three packets, x, y, and z. Since this routing solution delivers three distinct packets in two time slots, we say that it achieves a *multicast throughput* of 1.5 packets per channel use. (Here we think of a single "use" of the network "channel" as the transmission of at

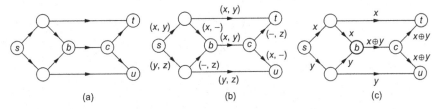

Figure 1.1 (a) the butterfly network, containing "bottleneck link" (b,c); (b) a "routing solution" to the multicast problem, showing three packets, x, y, and z routed to the destinations over two time slots ("–" indicates that the link is idle during the given time slot); (c) a "network coding" solution: the modulo-two sum $x \oplus y$ (XOR) of x and y is transmitted through the bottleneck, allowing both x and y to be decoded at each destination, thereby delivering two packets per time slot.

most one packet per network edge.) It can be shown that this multicast throughput is the best that can be achieved by any routing solution.

A "network coding solution"—achieving an improved multicast throughput of 2 packets per channel use—is shown in Fig. 1.1(c). Here the source again emits two packets x and y, but instead of routing one (and blocking the other), node b transmits their modulo-two sum $x \oplus y$ (i.e., their bit-by-bit exclusive OR). Node t now receives x and $x \oplus y$, and since $x \oplus (x \oplus y) = y$, node t can recover both x and y. Similarly, node u can recover both x and y from the two packets it receives. At the expense of an encoding operation at an internal node of the network, and decoding operations at the destination nodes, the multicast throughput has improved beyond what can be achieved by routing alone.

Can we do even better? In this case, it is easy to see that the answer is no, since nodes t and u are connected to the network by just two edges, and so can receive at most two pieces of information per unit time. (In general, as we will discuss later, the multicast throughput of a network is determined—as a consequence of the "max-flow min-cut" theorem—by the smallest "edge cut" separating the source and any of the destinations.) The network coding solution of Fig. 1.1(c) is, therefore, in fact *optimal* in the sense of achieving the greatest possible multicast throughput.

Another example—sometimes referred to as the "wireless butterfly network"—arises in the context of wireless relay networks. Consider the configuration illustrated in Fig. 1.2. The two valley stations, s and t, cannot communicate directly; instead they must communicate (over a radio

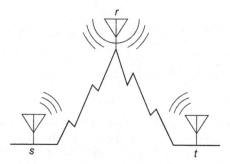

Figure 1.2 The "wireless butterfly," a wireless half-duplex relay network permitting two valley stations s and t to communicate via a (broadcasting) relay node r.

channel) through the mountain-top relay node r. We assume here that all radio equipment is limited to "half-duplex" operation, i.e., during any given interval, a node may either transmit or receive (but it may not transmit and receive simultaneously). During a given time period, the relay node may receive transmissions from either valley station (but not both), but when the relay broadcasts, both valley stations receive the message. Suppose that the communication objective is to communicate packets of m bits, sending packet x from s to t and packet y from t to s.

A "routing solution" might achieve this communication objective in four time slots: (1) send x from s to r; (2) broadcast x from r to t; (3) send y from t to r; and (4) broadcast y from r to s.

A "network coding solution" achieves this same communication objective in just three time slots: (1) send x from s to r; (2) send y from t to r; and (3) broadcast $x \oplus y$ from r to both s and t simultaneously. Since s already knows x, it can recover y by computing $x \oplus (x \oplus y)$, and similarly t can recover x. Again, at the expense of an encoding operation at an internal node of the network, and decoding operations at the destination nodes, the throughput has improved beyond what can be achieved by routing alone.

These two examples illustrate an important lesson: namely, that routing alone is in general insufficient to fully exploit the information-carrying capability of communication networks. Even though textbook descriptions of packet networks often treat "packet flow" as "commodity flow," the butterfly network shows that this analogy is not fundamentally accurate. Unlike cars on the highway, information packets can in fact be usefully combined in ways (for example, "linear superposition") that still allow the destinations to infer the messages being transmitted. This chapter, and indeed this entire book, is devoted to exploring the consequences of this simple idea.

The remainder of this chapter is organized as follows. Section 2 provides a basic graph-theoretic model of network communications, with an emphasis on the multicast case. The main theoretical result, described in Section 3.1, stems from the max-flow min-cut theorem: the multicast throughput is bounded by the least-capacity cut separating the source from any of the destinations. In Section 3.2, we describe *linear* network coding, where local coding operations are constrained to be linear over a finite field. In Section 3.3 we show that the upper bound on multicast throughput can be achieved by linear network coding over a finite field of sufficiently large size. In Section 4.1 we describe an efficient algorithm that, given a network, will provide linear network coding coefficients to solve a given multicast problem. In Section 4.2, we discuss random linear network coding, where these coding coefficients are not carefully designed, but simply chosen at random from a field. When the field size is sufficiently large, with high probability the randomly chosen code achieves the same performance as a carefully designed optimal code. In Section 5 we demonstrate that the multicast throughput advantage of coding over routing can be arbitrarily large. In Section 6 we discuss noncoherent random network coding, where neither the source nor any of the sinks has *a priori* knowledge of the specific linear coding operations performed by the network. In Section 7 we consider the nature of the alphabets needed to perform network coding, and present examples of flow problems where linear network coding is insufficient to achieve the maximum transmission rate.

The literature on network coding is vast and expanding rapidly. In addition to this chapter, other introductory treatments of network coding include [2–5].

Notation

Throughout this chapter, the number of elements in a finite set X will be denoted as $|X|$. Matrices will be denoted using a bold font, e.g., $\mathbf{A}$. The identity matrix is denoted as $\mathbf{I}$. A unit vector $(0, \ldots, 0, 1, 0, \ldots, 0)$, with a single one in position i, is denoted as $\mathbf{e}_i$.

2. GRAPHS AND NETWORKS

2.1. Combinational Packet Networks

The following graph-theoretical model, while not the most general possible, will be sufficient to model the main ideas of network coding.

A *combinational packet network* $\mathcal{N} = (V, E, S, T, A)$ comprises:

1. a finite directed acyclic multigraph $G = (V, E)$ where V is the set of vertices and E is the multiset of directed edges;
2. a distinguished set $S \subset V$ of *sources*;
3. a distinguished set $T \subset V$ of *sinks*;
4. and a finite packet alphabet A with $|A| \geq 2$.

Vertices model communication nodes within the packet network, while directed edges model error-free communication channels between the nodes. An edge (u, v) has *unit capacity* in the sense that it can be used to reliably deliver one *packet* (i.e., a symbol drawn from A) from u to v. To allow for greater (but still integral) capacity from u to v, parallel edges between u and v are permitted, i.e., G is in general a multigraph. To resolve any ambiguity when referring to one of several parallel edges, we will assume that the edges of multiset E are numbered as $e_1, e_2, \ldots, e_{|E|}$, where each e_i is an ordered pair of vertices with the possibility that $e_i = e_j$ for some $j \neq i$.

At vertex $v \in V$, we denote by $I(v)$ the set of (incoming) edges incident to v and by $O(v)$ the set of (outgoing) edges incident from v. Possibly by introducing "virtual nodes" in a given network we may, without loss of generality, assume that $S \cap T = \emptyset$, that $I(s) = \emptyset$ for every source node $s \in S$, and that $O(t) = \emptyset$ for every sink node $t \in T$.

As usual in graph theory, a (directed) *path* in G from a vertex $u \in V$ to a vertex $v \in V$ is a finite sequence of directed edges (u, v_1), $(v_1, v_2), (v_2, v_3), \ldots, (v_{\ell-2}, v_{\ell-1}), (v_{\ell-1}, v)$, where each element of the sequence is an element of E. A cycle in G is a directed path from some vertex to itself. The term *acyclic* in the definition of a packet network means that the multigraph G contains no cycles.

A vertex $v \in V$ is said to be *reachable* from $u \in V$ if $v = u$ or if there is a directed path in G from u to v. Reachability induces a partial order on V; this partial order can always be extended (not necessarily uniquely) to a compatible total order, called a *topological order*. Topological orders will be useful in Section 4.1.

An edge $(u, v) \in E$ takes in any packet $p \in A$ from u, and delivers that packet (without errors) to v. It is also possible that an edge remains idle (no packet is transmitted), and, in this case, it is assumed that v is aware that the edge is idle. (Equivalently, idle edges can be removed from E.) In the special case of linear network coding—discussed later—such "idle edges" can be indicated by transmission of the zero packet.

As in combinational logic (hence the terminology *combinational packet network*), it is assumed that packets transmitted on the non-idle edges of $O(v)$ are functions of the packets received on the non-idle edges of $I(v)$, or, if v is a source, of packets internally generated at v. In an actual network implementation, packets must be received (and possibly buffered) at v *before* outgoing packets can be computed and transmitted. Delays caused by transmission, buffering, and processing of packets are, however, not modelled explicitly.

We refer to the function applied by a particular node v as the *local encoding function* at v. For example, if v were acting as a router, then each packet sent on an edge of $O(v)$ would be a copy of one of the packets received on an edge of $I(v)$. The entire premise of network coding is that it may be useful to allow local encoding operations that are more general than simply copying packets.

By a single *channel use* of a combinational packet network $\mathcal{N}$ we mean a particular assignment of packets drawn from A to each of the (non-idle) edges of E, consistent with the particular local encoding functions implemented by the nodes of the network. In other words, in a single channel use, the local encoding functions are fixed, and each edge of the network is used at most once. Of course it is possible, as in the routing example of Fig. 1.1(b), that the local encoding functions may vary from channel use to channel use.

Throughout this chapter, transmission efficiencies will be measured in units of packets per channel use. For example, as previously noted, the butterfly network of Fig. 1.1(b) achieves a multicast rate of 1.5 packets per channel use, and that of Fig. 1.1(c) achieves a multicast rate of 2 packets per channel use. It should be noted that the measure of information ("packets") used here depends on the alphabet size $|A|$. It is straightforward to convert a rate from packets per channel use to bits per channel use, simply by multiplying by $\log_2 |A|$ bits per packet. The distinction between capacity measured in bits per channel use and capacity measured in packets per channel use becomes significant only when comparing networks with different packet alphabets, for example, in situations where the alphabet size is permitted to vary.

This combinational model is not the most general possible. For example, if the underlying transmission channels were wireless, then the network model would need to be modified to take into account the broadcast nature of the channel (i.e., the possibility that a single transmission might be received simultaneously by multiple receivers), and the multiple-access nature of the channel (i.e., that simultaneous transmissions from multiple transmitters may interfere with each other at a given receiver). The model also does not explicitly incorporate outage (the possibility that links may not be perfectly reliable), delay (or any notion of timing as in sequential logic), or the possibility of feedback (cycles in the graph). Nevertheless, the combinational model given is sufficiently rich that it will well illustrate the main ideas of network coding.

2.2. Network Information Flow Problems

Let the set of source vertices be given as $S = \{s_1, s_2, \ldots, s_{|S|}\}$. We assume that each source vertex $s_i \in S$ has an infinite supply of packets to send. We also assume that each sink vertex $t \in T$ is interested in reconstructing the packets sent by some subset $D_t \subseteq S$ of sources. We refer to the elements of D_t as the sources *demanded* by sink t.

Suppose that a source s_i has packets $p_1, p_2, p_3, \ldots$ to send. If, over n channel uses, a sink t can reconstruct (as a function of the packets it receives on $I(t)$ over those channel uses) k source packets $p_1, p_2, \ldots, p_k$, then we say that sink t can decode source s_i at rate $r_i = k/n$ packets per channel use. We will require that each sink demanding a particular source $s_i \in S$ be served at the same transmission rate r_i (packets per channel use). A particular collection of demands $\{D_t : t \in T\}$ is said to be *achievable* with rates $(r_1, r_2, \ldots, r_{|S|})$ over a given network if, for some choice of (possibly time-varying) local encoding operations, and some n, each sink can reconstruct the sources it demands at the specified rate.

An important special case of this general network information flow problem is the *multicast problem*, in which $D_t = S$ for all $t \in T$, i.e., the case where all sources are demanded by all sinks.

Other important special cases include the *unicast problem* (with a single source s, a single sink t, and $D_t = \{s\}$) or the problem of *multiple unicasts* (for some integer $L > 1$, $S = \{s_1, \ldots, s_L\}$, $T = \{t_1, \ldots, t_L\}$, and $D_{t_i} = \{s_i\}$). Of course many further variations are possible.

In the following sections, we will consider first the problem of a single-source multicast problem, and show that linear network coding achieves the maximum possible multicast rate, provided that the packet alphabet is a sufficiently large finite field. However, we will see that linear network coding is not in general sufficient to achieve optimal rates in general network information flow problems.

3. THE SINGLE-SOURCE MULTICAST PROBLEM

3.1. Multicast Capacity

Suppose that vertex $t \in V$ is reachable in G from vertex $s \in V$. An $\{s, t\}$-separating cut is a collection C of edges such that every path from s to t contains an edge from C. If the edges of C were removed from E, then there would be no path from s to t. A *minimal cut* separating s and t is a cut of the smallest possible cardinality (among all $\{s, t\}$-separating cuts). We denote the cardinality of an $\{s, t\}$-separating minimal cut as mincut(s, t).

A minimal cut can be regarded as a bottleneck between s and t, as every path from s to t passes through the bottleneck. It is intuitively obvious that such bottlenecks must limit the rate of information flow between s and t. This statement can be made more precise as follows.

For any set $B \subseteq E$ of edges, ordered as in E, let X_B denote the collection of packets transmitted over the non-idle edges of B. Now let s be a source, t a sink, and C any $\{s, t\}$-separating cut. Note that $X_{O(s)} \to X_C \to X_{I(t)}$ is a Markov chain. It follows from the data-processing inequality of information theory (see, e.g., [6]) that $I(X_{O(s)}; X_{I(t)}) \leq I(X_{O(s)}; X_C)$, i.e., the mutual information between packets sent by s and packets received by t cannot exceed the mutual information between packets sent by s and those that pass through the cut C. A trivial upper bound on $I(X_{O(s)}; X_C)$ is $|C|$ packets per channel use; thus:

$$I\left(X_{O(s)}; X_{I(t)}\right) \leq |C|,$$

where C is *any* $\{s, t\}$-separating cut. Minimizing $|C|$ over all cuts yields the upper bound:

$$I\left(X_{O(s)}; X_{I(t)}\right) \leq \text{mincut}(s, t)$$

on the rate of information flow between s and t. In words, if $R(s, t)$ denotes the communication rate between node s and node t, then:

$$R(s, t) \leq \text{mincut}(s, t).$$

To achieve the upper bound, it is necessary, for each minimal cut C, that $I(X_{O(s)}; X_C) = |C|$, which means that packets sent through a minimal cut must be independent and uniformly distributed over A; furthermore, it is necessary that the original source packets be decodable from those that flow over any cut C (including $O(s)$).

It is well known from the edge-connectivity version of Menger's Theorem [7], or from the theory of commodity flows [8, 9], that mincut(s, t) is equal to the maximum number of pairwise edge-disjoint paths from s to t. Such a collection of pairwise edge-disjoint paths can be found using, e.g., the Ford-Fulkerson algorithm. Transmission of $R(s, t) = \text{mincut}(s, t)$ packets per channel use from the source s to the single sink t can therefore be achieved by *routing*, sending one packet along each of the edge-disjoint paths between s and t in each channel use.

Suppose now, that the set of sinks, T, contains more than one node. It is clear that the multicast rate $R(s, T)$ from s to T cannot exceed the transmission rate that can be achieved from s to any element of T, i.e., the multicast rate $R(s, T)$ must satisfy:

$$R(s, T) \leq \min_{t \in T} \text{mincut}(s, t). \tag{1}$$

The main theorem of network multicasting is that the upper bound (1) is achievable (with equality) via network coding [1]. Moreover, the upper bound is achievable with *linear* network coding, provided that the packet alphabet A is a sufficiently large finite field [10]. The quantity $\min_{t \in T} \text{mincut}(s, t)$ is referred to as the *multicast capacity* of the given combinational packet network.

3.2. Linear Network Coding

Before formally stating the theorem, let us define what is meant by "linear network coding over a finite field." In linear network coding, the packet alphabet A is a q-element finite field F_q (or, more generally, a space of row vectors over F_q). For readers unfamiliar with finite fields, Appendix A gives a brief primer. For now, it is only necessary to appreciate that finite fields endow a finite set with well-defined and efficiently-implementable addition, subtraction, multiplication, and division operations, allowing for the solution of systems of simultaneous linear equations using all of the tools—matrices, determinants, Gauss-Jordan elimination, etc.—familiar from linear algebra. The smallest possible finite field is $F_2 = \{0, 1\}$ under integer arithmetic modulo 2.

By a slight abuse of notation, we assume that the message packets are given as $(p_1, p_2, \ldots, p_r)^T$, regarded as an "incoming" column vector $X_{I(s)}$ at the source node s. Throughout this section, we will assume that each packet p_i is a scalar from F_q. The extension to the more general case where the packets are vectors of length m over F_q is immediate; for example, $X_{I(s)}$ would be replaced with an $r \times m$ matrix over F_q, in which the ith row is the packet p_i. In the butterfly network example of Section 1, we had $q = 2$ with a packet length of m bits.

Local encoding functions are *linear* over F_q, i.e., for each internal vertex v, the (column) vector of transmitted packets $X_{O(v)}$ is related to the (column) vector of received packets $X_{I(v)}$ by the linear equation:

$$X_{O(v)} = \mathbf{L}_v X_{I(v)}$$

where $\mathbf{L}_v$ is a matrix of coefficients from F_q, called the local transfer matrix at v. In other words, each outgoing packet (a component of $X_{O(v)}$) is an F_q-linear combination of the incoming packets at node v (the components of $X_{I(v)}$). The row of $\mathbf{L}_v$ corresponding to a particular edge $e \in O(v)$ is called the *local encoding vector* associated with e.

Note that, since only linear operations are performed in the network, every packet transmitted along an edge is a linear combination of the source packets $p_1, p_2, \ldots, p_r$. In particular, for every vertex $v \in V$, we have:

$$X_{I(v)} = \mathbf{G}_v \begin{bmatrix} p_1 \\ \vdots \\ p_r \end{bmatrix}$$

where again $\mathbf{G}_v$ is a matrix of coefficients from F_q, called the *global transfer matrix* at v. The row of $\mathbf{G}_v$ corresponding to a particular edge $e \in I(v)$ is called the *global encoding vector* associated with e.

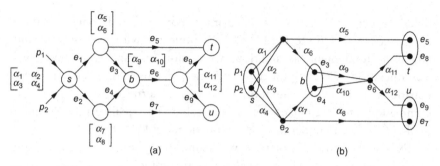

Figure 1.3 The butterfly network (a) labeled with local transfer matrices; and (b) the corresponding line graph.

Yet another useful way to describe a combinational packet network under linear network coding is via a state-space description. For simplicity, let us remove all idle edges from E. We introduce the so-called *line graph* (e.g., [11],[12, p. 168]) of the network, the directed graph $G' = (V', E')$, where $V' = E$ (i.e., the vertices of G' correspond to the edges of G), and with $(e_i, e_j) \in E'$ if, and only if, we have $e_i \in I(v)$ and $e_j \in O(v)$ for some $v \in V$. For example, Fig. 1.3(b) illustrates the line graph corresponding to the augmented butterfly network of Fig. 1.3(a).

Let X be an $|E| \times 1$ vector whose ith component is the packet sent on edge e_i of the network. We view X as the network *state*. A standard state-space model for the linear combinational packet network is given (at sink node t) as:

$$X = \mathbf{A}X + \mathbf{B}U$$

$$X_{I(t)} = \mathbf{C}_t X$$

where the $|E| \times |E|$ matrix $\mathbf{A}$ is the edge-to-edge transfer matrix (whose (i,j)th entry gives the transmittance coefficient from edge e_i to edge e_j), $U = (p_1, p_2, \ldots, p_r)^T$ is the vector of input packets, $\mathbf{B}$ is the $|E| \times r$ matrix giving the transmittance coefficient from each input packet to each edge of E, and $\mathbf{C}_t$ is an $|I(t)| \times |E|$ matrix projecting X onto those components observed at sink node t. Intuitively, matrix $\mathbf{B}$ specifies how the message packets couple to the edges of the network, matrix $\mathbf{A}$ specifies how the edges couple to each other, and matrix $\mathbf{C}_t$ specifies how the edges couple to sink node t. It is easy to see that the global transfer matrix $\mathbf{G}_t$ at sink node t can be obtained as:

$$\mathbf{G}_t = \mathbf{C}_t(\mathbf{I} - \mathbf{A})^{-1}\mathbf{B}. \qquad (2)$$

For any k, the entries of $\mathbf{A}^k$ represent the coupling coefficients between edges due to paths of length k in the line graph G'. Since the network is finite and acyclic, we have $\mathbf{A}^L = 0$ for some sufficiently large L, i.e., the matrix $\mathbf{A}$ is nilpotent. Thus $(\mathbf{I} - \mathbf{A})^{-1} = \mathbf{I} + \mathbf{A} + \mathbf{A}^2 + \cdots + \mathbf{A}^{L-1}$ is a polynomial in $\mathbf{A}$.

To illustrate with a concrete example, using the local transfer matrices shown in Fig. 1.3(a), we have $U = (p_1, p_2)^T$, $\mathbf{A} = [a_{i,j}]$ with $a_{1,3} = \alpha_6$, $a_{1,5} = \alpha_5$ (and $a_{1,j} = 0$ if $j \notin \{3,5\}$), $a_{2,4} = \alpha_7$, $a_{2,7} = \alpha_8$ (and $a_{2,j} = 0$ if $j \notin \{4,7\}$), etc.:

$$\mathbf{B} = \begin{bmatrix} \alpha_1 & \alpha_2 \\ \alpha_3 & \alpha_4 \\ 0 & 0 \\ \vdots & \vdots \\ 0 & 0 \end{bmatrix}, \qquad \mathbf{C}_t = \begin{bmatrix} 0 & 0 & 0 & 0 & 1 & 0 & 0 & 0 & 0 \\ 0 & 0 & 0 & 0 & 0 & 0 & 0 & 1 & 0 \end{bmatrix}, \text{ and}$$

$$\mathbf{C}_u = \begin{bmatrix} 0 & 0 & 0 & 0 & 0 & 0 & 0 & 0 & 1 \\ 0 & 0 & 0 & 0 & 0 & 0 & 1 & 0 & 0 \end{bmatrix}.$$

Then, either from (2), or by treating Fig. 1.3(b) as a signal flow graph with p_1 and p_2 as inputs, we obtain:

$$\mathbf{G}_t = \begin{bmatrix} \alpha_1\alpha_5 & \alpha_2\alpha_5 \\ \alpha_1\alpha_6\alpha_9\alpha_{11} + \alpha_3\alpha_7\alpha_{10}\alpha_{11} & \alpha_2\alpha_6\alpha_9\alpha_{11} + \alpha_4\alpha_7\alpha_{10}\alpha_{11} \end{bmatrix} \text{ and}$$

$$\mathbf{G}_u = \begin{bmatrix} \alpha_1\alpha_6\alpha_9\alpha_{12} + \alpha_3\alpha_7\alpha_{10}\alpha_{12} & \alpha_2\alpha_6\alpha_9\alpha_{12} + \alpha_4\alpha_7\alpha_{10}\alpha_{12} \\ \alpha_3\alpha_8 & \alpha_4\alpha_8 \end{bmatrix}.$$

Note that the entries of $\mathbf{G}_t$ and $\mathbf{G}_u$ are polynomials in the αs, as expected.

3.3. Linear Network Coding Achieves Multicast Capacity

Returning now to the general case, in order for a sink node t to be able to recover every element of $(p_1, p_2, \ldots, p_r)$ it is necessary for the $|I(t)| \times r$ global transfer matrix $\mathbf{G}_t$ at t to have rank r, since only then does $\mathbf{G}_t$ have a left inverse $\mathbf{G}_t^{-1}$ satisfying $\mathbf{G}_t^{-1}\mathbf{G}_t = \mathbf{I}_r$, where $\mathbf{I}_r$ is the $r \times r$ identity matrix. (Note that $\mathbf{G}_t$ does not need to be *square* to have a left inverse, merely full rank.) Forming $\mathbf{G}_t^{-1}X_{I(t)}$ at node t produces $(p_1, p_2, \ldots, p_r)^T$, as desired.

The proof of the following theorem shows that if $r \leq \min_{t \in T} \text{mincut}(s, t)$ and q is sufficiently large, then it is possible to make $\mathbf{G}_t$ an $r \times r$ invertible matrix at each sink node t, thereby achieving a multicast rate of r with linear network coding.

Theorem 1 (Linear Network Multicasting Theorem). *Let $\mathcal{N} = (V, E, \{s\}, T, F_q)$. A multicast rate of $R(s, T) = \min_{t \in T} \text{mincut}(s, t)$ is achievable, for sufficiently large q, with linear network coding.*

We provide a proof of this theorem following the elegant algebraic framework of [11], as the proof technique will be also be useful in other chapters of this book. A main tool is the so-called "Sparse Zeros Lemma," which is a consequence of a Lemma proved in Appendix B. Recall that $F_q[x_1, \ldots, x_n]$ denotes the set of polynomials in n variables $x_1, \ldots, x_n$ with coefficients from the field F_q.

Lemma 1 (Sparse Zeros Lemma). *Let $f \in F_q[x_1, \ldots, x_n]$ be a polynomial, not identically zero, whose x_i-degree is at most d for all i. If $q > d$, then $f(a_1, \ldots, a_n) \neq 0$ for some $(a_1, \ldots, a_n) \in F_q^n$.*

Proof of Theorem 1: Let $r = \min_{t \in T} \text{mincut}(s, t)$, and let $p_1, p_2, \ldots, p_r$ denote the source packets that we wish to transmit. For each sink node t it is possible to find r edge-disjoint paths from s to t. We delete from E (or set to idle) any edges not on any of these paths, and refer to the network containing only non-idle edges as the *reduced network*. Note that $|I(t)| = r$ for all sink nodes $t \in T$ in the reduced network, which makes each global transfer matrix $\mathbf{G}_t$ an $r \times r$ matrix.

Following [11], we introduce an indeterminate (variable) for each entry of each local transfer matrix in the network. Let us denote these variables as $\alpha_1, \alpha_2, \ldots, \alpha_\eta$, as illustrated (for the butterfly network) in Fig. 1.3(a). As noted, since G is acyclic, each global transfer matrix $\mathbf{G}_t(\alpha_1, \ldots, \alpha_\eta)$, $t \in T$, has polynomial entries. Our task is to demonstrate that each matrix $\mathbf{G}_t$ can be made invertible for some specific choice of values for the indeterminates from a sufficiently large finite field.

Now clearly $\mathbf{G}_t$ is invertible if, and only if, $\det(\mathbf{G}_t) \neq 0$. In fact $\det(\mathbf{G}_t)$ is a (multivariate) polynomial in the indeterminate variables. This polynomial is not identically the zero polynomial, since it is always possible to set the value of the indeterminate variables to achieve the routing solution from s to t, which yields the unit determinant matrix $\mathbf{G}_t = \mathbf{I}_r$.

Because the network is acyclic, we note that any variable α_i may in general appear with degree at most one in any of the entries of $\mathbf{G}_t$. Since $\mathbf{G}_t$ is $r \times r$, $\det(\mathbf{G}_t)$ clearly is a polynomial with degree bounded by r, i.e., for any i, the maximum α_i-degree of the determinant is at most r. A somewhat surprising observation (proved in Appendix C) is that $\det(\mathbf{G}_t)$ is in fact a multivariate polynomial with degree bounded by one!

Now to complete the proof, let:

$$f(\alpha_1, \ldots, \alpha_\eta) = \prod_{t \in T} \det \left(\mathbf{G}_t \left(\alpha_1, \ldots, \alpha_\eta \right) \right).$$

Since f is a product of polynomials of degree bounded by one (with coefficients from a field), none of which is identically the zero polynomial, it follows that f is a polynomial of degree bounded by $|T|$, not identically the zero polynomial. From the Sparse Zeros Lemma, it follows that as long as $q > |T|$, the indeterminate variables can be chosen from F_q to satisfy $f \neq 0$, thereby making $\det(\mathbf{G}_t) \neq 0$ for all $t \in T$. This proves the theorem. ∎

The linear network multicasting theorem can be extended to the case of multicasting from multiple sources ($|S| > 1$) by augmenting the network with a virtual "supersource" connected to each actual source via an appropriate number of parallel edges. Specifically, let $S = \{s_1, \ldots, s_{|S|}\}$ and suppose that sink node t is reachable from these source nodes with a number of edge-disjoint paths, in particular with $r(s_i, t)$ edge-disjoint paths from s_i to t. Clearly source s_i can route packets to t at a rate of $r(s_i, t)$ packets per channel use. Now let $r_i = \min_{t \in T} r(s_i, t)$, and let $r = r_1 + r_2 + \cdots + r_{|S|}$. Augment the network with a "virtual source" s^*, connected to s_i with r_i parallel edges. Note that each sink $t \in T$ is reachable from s^* by r edge-disjoint paths. Theorem 1 implies that s^* can, using linear network coding, multicast to all sinks at a rate of r packets per channel use. Let us fix one such linear network coding solution. We then have, for some $r \times r$ matrix $\mathbf{L}_{s^*}$:

$$X_{O(s^*)} = \mathbf{L}_{s^*}(p_1, \ldots, p_r)^T,$$

i.e., the packets transmitted on the edges incident from s^* are linear combinations of the r supersource packets. However, since $O(s^*)$ is a cut separating s^* from the rest of the network, $(p_1, \ldots, p_r)$ must be decodable from $X_{O(s^*)}$, which means that $\mathbf{L}_{s^*}$ must be invertible. Premultiplying at s^*

by $\mathbf{L}_{s*}^{-1}$ does not affect the ability of any of the sinks to decode, but translates a general network coding solution into a solution where the global encoding vectors on the edges of $O(s^*)$ are unit vectors. Thus, source s_i receives r_i "uncoded" supersource packets. By supposing that these r_i supersource packets are actually generated at s_i, we see that the given linear network coding solution has the effect of delivering packets from s_i to each sink at a rate r_i packets per channel use, thereby achieving the rate-tuple $(r_1, \ldots, r_{|S|})$.

Theorem 1 has another sometimes useful corollary. Let $\mathcal{N} = (V, E, \{s\}, T, F_q)$. When q is sufficiently large, a linear network coding solution exists for which $\mathbf{G}_t$ has rank mincut(s, t) for all $t \in T$. In other words, the global transfer matrices associated with the sinks can simultaneously be made to achieve their maximum possible rank. To see this, let $r = \max_{t \in T}$ mincut(s, t). For each sink $t \in T$ augment the network with $r -$ mincut(s, t) parallel edges from s to t. According to Theorem 1, linear network coding can be used in the augmented network to achieve a multicast rate of r. This linear network coding solution induces a rank r global transfer matrix at each sink node $t \in T$. If the augmenting edges at each sink node t are now deleted, the corresponding rows of the global transfer matrix are deleted, resulting in a global transfer matrix $\mathbf{G}_t$ having rank mincut(s, t).

4. CONSTRUCTION OF NETWORK CODES FOR MULTICASTING

4.1. The Linear Information Flow Algorithm

Theorem 1 shows that the single-source multicast capacity of a network can be achieved via linear network coding over a finite field F_q, with $q > |T|$. Given a network, is it possible to find such a linear network coding solution in a computationally efficient manner? Fortunately the answer is yes; in this section we briefly sketch the "linear information flow" algorithm of [13] that provides a polynomial–time solution to this problem.

As in the proof of Theorem 1, we let $r = \min_{t \in T}$ mincut(s, t), and let $p_1, p_2, \ldots, p_r$ denote the source packets that we wish to transmit. We will assume that $|T| \geq 2$, and that a finite field F_q with $q \geq |T|$ is fixed. Note that equality $q = |T|$ is permitted. The algorithm proceeds in two phases.

In the first phase of the algorithm, use, say, the Ford-Fulkerson algorithm for each $t \in T$, to find r edge-disjoint paths $p_1(t), \ldots, p_r(t)$ from s to t. Only the edges in these paths will be needed; therefore we delete from E (or set to idle) any edges not on any of these paths, and refer to the network containing only non-idle edges as the *reduced network*. For each edge e in the reduced network, and for each sink $t \in T$, and for $i \in \{1, \ldots, r\}$, define the indicator function:

$$\delta(e, t, i) = \begin{cases} 1 & \text{if } e \in p_i(t); \\ 0 & \text{otherwise.} \end{cases}$$

The function $\delta(e, t, i)$ indicates whether edge e lies on path $p_i(t)$ from s to t. For fixed e and t, since the r chosen paths from s to t are edge-disjoint, $\delta(e, t, i) = 1$ for at most one value of i. Edge e is said to *affect* a sink $t \in T$ if $\delta(e, t, i) = 1$ for some $i \in \{1, \ldots, r\}$.

In the second phase of the algorithm, a $1 \times r$ global encoding vector is chosen (as described below) for each of the edges of the reduced network. The edges are processed in topological order, i.e., if a vertex v_i precedes v_j in topological order, then the edges of $O(v_i)$ are processed prior to those of $O(v_j)$. This ordering ensures that the coding vectors for predecessor (i.e., "upstream") edges are selected prior to selecting the coding vectors for a given edge. In particular, if the directed path $p_i(t)$ comprises edges $e_{i_1}, e_{i_2}, \ldots, e_{i_L}$ (in that order), then the global encoding vector for e_{i_1} is chosen before that of e_{i_2}, which is chosen before that of e_{i_3}, and so on. Thus, at any stage of the algorithm, each path $p_i(t)$ includes a well-defined *most recently updated* edge, i.e., the edge e_{i_j} assigned a global encoding vector, whose successor $e_{i_{j+1}}$ has *not* yet been assigned a global encoding vector. We will denote the global encoding vector assigned to the most recently updated edge of $p_i(t)$ as $\mathbf{h}_i(t)$. Subsequent to initialization, the central invariant maintained by the algorithm is that, for every sink $t \in T$, $\{\mathbf{h}_1(t), \ldots, \mathbf{h}_r(t)\}$ is a basis for F_q^r; or, equivalently, that the $r \times r$ matrix:

$$\mathbf{H}_t = \begin{bmatrix} \mathbf{h}_1(t) \\ \vdots \\ \mathbf{h}_r(t) \end{bmatrix}$$

is nonsingular. Upon termination of the algorithm, the most recently updated edges for each sink t are precisely the incoming edges of $I(t)$, and the matrix $\mathbf{H}_t$ is equal to the global transfer matrix $\mathbf{G}_t$.

Initialization in the second phase of the algorithm amounts to selection of the global encoding vectors for the edges of $O(s)$. Each sink node t maintains a set H_t (initially empty) of the most recently updated global encoding vectors associated with $p_1(t), \ldots, p_r(t)$. When a global encoding vector $\mathbf{h}(e)$ is chosen for an edge $e \in O(s)$, the vector $\mathbf{h}(e)$ must be chosen so as to be linearly independent of the vectors in H_t for all $t \in T$ affected by e (i.e., for those $t \in T$ for which $\delta(e, t, i) = 1$ for some i). Now $|H_t| < r$ (i.e., if $e \in O(s)$ affects t, then H_t must so far have been assigned fewer than r vectors), and so each H_t rules out the zero vector and at most $q^{r-1} - 1$ nonzero vectors as potential values for $\mathbf{h}(e)$. In total, among the q^r vectors in F_q^r, at most $1 + |T|(q^{r-1} - 1)$ vectors are ruled out, leaving at least:

$$q^{r-1}(q - |T|) + |T| - 1$$

vectors to choose from. As long as $q \geq |T|$, this number is strictly positive, and thus the initialization procedure is guaranteed to succeed in assigning global encoding vectors so that each set H_t is a basis for F_q^r, or equivalently so that the matrix $\mathbf{H}_t$ is nonsingular.

With initialization now complete, let us consider the processing of an edge e, as illustrated in Fig. 1.4. Assignment of a local encoding vector at edge e will have the effect of replacing (at most) one global encoding vector at each sink—these are shown by the shaded boxes in Fig. 1.4. Each of the shaded boxes will be replaced by the same vector, which itself must be a linear combination of the incoming global encoding vectors illustrated in the figure. Suppose the global encoding vectors associated with edges incoming to e span an m-dimensional vector space V_1. At sink node t, the vectors *not* being replaced span an $(r - 1)$-dimensional vector space V_2. The dimension of $V_1 \cap V_2$ is given as:

$$\dim(V_1 \cap V_2) = \dim(V_1 + V_2) - \dim(V_1) - \dim(V_2)$$
$$= r - m - (r - 1) = m - 1.$$

(By construction, $\dim(V_1 + V_2) = r$ since V_1 must contain the vector being replaced.) The space $V_1 \cap V_2$ contains the zero vector and $q^{m-1} - 1$

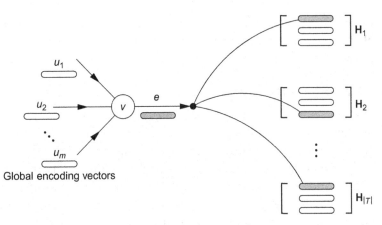

Figure 1.4 The key step in the linear information flow algorithm: the coding vector for edge e must be chosen as a linear combination of "incoming" global encoding vectors $u_1, \ldots, u_m$ so that $\mathbf{H}_t$ remains invertible for each sink $t \in T$.

nonzero vectors, all of which are ruled out as permissible choices of global coding vector for edge e. In total, among the q^m vectors in V_1, at most $1 + |T|(q^{m-1} - 1)$ vectors are ruled out, leaving at least:

$$q^{m-1}(q - |T|) + |T| - 1$$

vectors to choose from. As long as $q \geq |T|$, this number is strictly positive, and hence the algorithm is guaranteed to succeed.

Note that if the local encoding vector for edge e is simply chosen at random from V_1, then, given that previous choices have been successful, the probability of success, i.e., the probability of choosing a vector—in a single trial—that maintains the rank of each matrix $\mathbf{H}_t$, is at least $1 - |T|/q$. Such a random selection is accomplished simply by selecting the relevant local encoding coefficients independently and uniformly at random from F_q.

In [13] it is shown that the linear information flow algorithm can be efficiently implemented in the sense that it's running time scales as a polynomial in $|E|$, $|T|$, and $\log(q)$. A key efficiency is obtained by realizing that testing a vector of length r for membership in an $(r-1)$-dimensional vector space is equivalent to testing a vector for membership in an $(r, r-1)$ linear code; such a code is described by a single parity-check equation (equivalently, the dual code is 1-dimensional), and hence the test can be performed via a single inner-product operation.

An alternative to the linear information flow algorithm is a matrix-completion approach due to Harvey, *et al.* [14]. Barbero and Ytrehus [15] provide heuristics for achieving a small field size.

4.2. Random Construction

Suppose that, instead of carefully designing a linear network code using, say, the linear information flow algorithm, the local encoding vector is simply chosen completely at random. What is the probability that such a randomly chosen linear network code will achieve the multicast capacity of the network? This "random approach" was investigated by Ho, *et al.* [16], and, as we will discuss in Section 6, the use of random linear network coding has many benefits in practical applications.

Let $\eta \leq |E|$ denote the number of "random coding edges," i.e., edges at which the local encoding vectors are chosen randomly, and consider the operation of the linear information flow algorithm described in Section 4.1. Suppose that at each step of the algorithm involving a random coding edge, the local encoding coefficients are chosen independently and uniformly at random from F_q. (On the non-random coding edges, we assume that the local encoding coefficients are chosen so that each $\mathbf{H}_t$ matrix surely maintains its full rank.) As noted above, the probability of success (i.e., the probability that the induced global coding vector will maintain the full rank of each $\mathbf{H}_t$ matrix) in the first random coding edge processed is at least $1 - |T|/q$. Given that the first outcome is successful, the probability of success in processing the second random coding edge is also at least $1 - |T|/q$; thus the probability that *both* trials are successful is at least $(1 - |T|/q)^2$. Continuing this process, and noting that the probability of success in the ith trial, given success in all previous trials, is at least $1 - |T|/q$, we find that the probability that all η trials are successful is at least:

$$P[\text{success}] \geq (1 - |T|/q)^\eta \geq 1 - \frac{\eta|T|}{q}.$$

To make $P[\text{success}] \geq 1 - \delta$ (for some small $\delta > 0$) it suffices that:

$$q \geq \frac{|T|\eta}{\delta}.$$

Thus we observe that, when the field size q is sufficiently large, a randomly chosen network code will, with high probability, achieve the multicast capacity. In practice, a field size of $q = 2^m$ is preferred, where m is the number of bits per field element. We see that the probability that, under random linear network coding, the probability that the code fails to achieve the multicast capacity satisfies:

$$P[\text{failure}] \leq |T|\eta 2^{-m},$$

which decreases exponentially with the number of bits per field element.

A more detailed analysis given in [16] can be used to strengthen this result. In some cases (for example, if $|I(v)| = 1$) it may be possible for a node v to determine the local encoding vector for an outgoing edge deterministically, with no reduction in the achievable multicast rate. We call an edge a "random coding edge" if its local encoding vector is not fixed deterministically, but instead is chosen randomly. Let c_t denote the number of random coding edges encountered on the r edge-disjoint paths from the source s to a sink node t constructed in the first phase of the linear information flow algorithm, and let $\eta' = \max_{t \in T} c_t$. Then the probability that randomly chosen local encoding coefficients result in a feasible network code is at least:

$$P[\text{success}] \geq (1 - |T|/q)^{\eta'}.$$

A key result needed to prove this bound is a refinement of the Schwartz-Zippel Theorem, described in Appendix B.

5. CODING VERSUS ROUTING

The following example, due to [13], shows that the gain of coding over routing can be arbitrarily large. Consider the combinational packet network shown in Fig. 1.5. Here there is a single source s, $2h$ intermediate nodes $u_1, \ldots, u_{2h}$, and $\binom{2h}{h}$ sinks $t_1, \ldots, t_{\binom{2h}{h}}$. Each sink is connected to a distinct subset of h of the intermediate nodes.

The multicast capacity of this network is h packets per channel use. This capacity can be achieved using a classical linear rate-$1/2$ maximum-distance-separable erasure-correcting code $\mathcal{C}$ of length $2h$, e.g., a (shortened) Reed-Solomon code over F_q, $q \geq 2h$ (see, e.g., [17]). Such a code is capable of correcting any pattern of h erasures in a block of $2h$ symbols. In

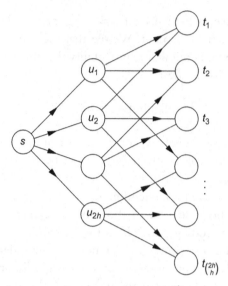

Figure 1.5 A network with $2h$ intermediate nodes and $\binom{2h}{h}$ sink nodes, each connecting to a different subset of intermediate nodes. This network has a multicast capacity of h; however, routing can achieve a multicast capacity of less than 2.

this application, the ith codeword symbol (which is a linear combination of the h source symbols) is sent to intermediate node u_i, which simply relays this to the sink nodes to which it is connected. Each sink sees exactly h out of the $2h$ symbols of the codeword, and therefore can decode using the erasure-correcting algorithm associated with $\mathcal{C}$. In this example, all of the "network coding" (i.e., forming linear combinations of source packets) actually occurs at the source, not at any of the intermediate nodes in the network.

Under routing, however, the multicast capacity is less than 2 packets per channel use, independent of h. To see this, suppose the source wishes to send at rate 2, routing $2n$ packets, $p_1, p_2, \ldots, p_{2n}$ in n channel uses to each of the sinks. Let U_i denote the subset of intermediate nodes that receive packet p_i. Since at most a total of $2nh$ packets can be distributed to the intermediate nodes, it must be the case that:

$$\sum_{i=1}^{2n} |U_i| \le 2nh,$$

and so (by the pigeonhole principle) $|U_j| \le h$ for some j. It follows that packet p_j is delivered to at most h intermediate nodes, and hence there is a set W of at least h intermediate nodes to which p_j is *not* delivered. There is at least one sink t_W that connects only to intermediate nodes in W, and t_W does not receive p_j. Thus a multicast rate of 2 (or more) packets per channel use cannot be achieved by routing.

In general, the best multicast rate that can be achieved by (fractional) routing corresponds to an optimal packing of multicast trees in the network; this is sometimes referred to as the "fractional Steiner tree-packing number." A multicast tree in G is a directed subtree of G that spans the source node and all the sink nodes. Let K denote the set of multicast trees and let K_e denote the subset of K containing those trees that use a link e. We will assign a flow rate of f_k to a tree $k \in K$, i.e., every edge of k is assigned a flow rate of f_k. The maximum rate that we can achieve is:

$$\max_{f_k : k \in K} \quad \sum_{k \in K} f_k$$
$$\text{subj. to :} \quad \sum_{k \in K_e} f_k \le 1 \text{ for all } e \in E,$$
$$f_k \ge 0 \text{ for all } k \in K.$$

In words, this linear program asks for an assignment of flow to each tree $k \in K$ so as to maximize the total flow, subject to the constraint that the total flow assigned to any (unit capacity) edge $e \in E$ cannot exceed unit flow, and the constraint that each flow must be non-negative.

As the number of multicast trees in G may be very large, it will generally be computationally infeasible to solve this linear program, except for very small networks. Indeed, it is known that optimal fractional packing of distribution trees is NP hard [18]. In contrast, as we have seen, multicast capacity-achieving network codes can be found in a computationally efficient manner (or even at random).

In [19], it is shown that the multicast coding advantage, i.e., the ratio of the multicast rate that can be achieved with network coding and the rate achieved by routing, is equal to the integrality gap of a linear programming relaxation for the minimum weight directed Steiner tree problem. Known lower bounds on this integrality gap are $\Omega((\log|V|/\log\log|V|)^2)$ [20] and $\Omega(\sqrt{|V|})$ [21]. Using the gadget introduced in [21], Chekuri *et al.* [22] have constructed networks in which the multicast coding advantage scales in proportion to $\sqrt{|T|}$. The example of Fig. 1.5 exhibits a multicast coding advantage of $\Omega(\log|T|)$.

6. NONCOHERENT NETWORK CODING

6.1. Transmission with Packet Headers

The random linear network coding approach can be used not only to design a fixed network code, but also as a *protocol* [16, 23].

In this approach, communication between source and sinks occurs in a series of rounds or "generations;" during each generation, the source injects a number of fixed-length packets into the network, each of which is a fixed length row vector over a finite field F_q. These packets propagate throughout the network, possibly passing through a number of intermediate nodes between transmitter and receiver. Whenever an intermediate node has an opportunity to send a packet, it creates a random F_q-linear combination of the packets it has available, and transmits this random combination. Finally, the receiver collects such randomly generated packets and tries to infer the set of packets injected into the network. This model is robust in the sense that we make no assumption that the network operates synchronously or without delay, or even that the network is acyclic.

The set of successful packet transmissions in a generation induces a directed multigraph with the same vertex set as the network, in which edges denote successful packet transmissions. As we have seen, the rate of information transmission (packets per generation) between the source and the sinks is upper-bounded by the smallest min–cut between these nodes, and, furthermore, random linear network coding in F_q is able to achieve a transmission rate equal to the min–cut rate with probability approaching one as $q \to \infty$ [16].

If we let the message packets form the rows of a matrix:

$$\mathbf{P} = \begin{bmatrix} p_1 \\ \vdots \\ p_r \end{bmatrix},$$

and assume that sink node t collects L packets $y_1, \ldots, y_L$ in a generation, forming the rows of a matrix $\mathbf{Y}$, then:

$$\mathbf{Y} = \mathbf{G}_t \mathbf{P}$$

where the $L \times r$ matrix $\mathbf{G}_t$ is the global transfer matrix induced at sink node t by the random choice of local encoding coefficients in the network.

If we assume that $\mathbf{G}_t$ is not known to either the source or the sink t, then we refer to the network operation as *noncoherent* network coding.

How can the sink recover $\mathbf{P}$? As in noncoherent multi-antenna communication, an obvious method is to endow the transmitted packets with so-called "headers" that can be used to record the particular linear combination of the components of the message present in each received packet. In other words, we let packet $p_i = (\mathbf{e}_i, m_i)$, where $\mathbf{e}_i$ is the ith unit vector and m_i is the packet "payload." We then have $\mathbf{P} = [\mathbf{I}_r \ \mathbf{M}]$, so that:

$$\mathbf{Y} = \mathbf{G}_t\mathbf{P} = \mathbf{G}_t\,[\mathbf{I}_r \ \mathbf{M}] = [\mathbf{G}_t \ \mathbf{G}_t\mathbf{M}].$$

The first r columns of $\mathbf{Y}$ thus yield $\mathbf{G}_t$, and, assuming $\mathbf{G}_t$ has a left inverse $\mathbf{G}_t^{-1}$ (an event that occurs with high probability if the field size is sufficiently large), sink node t can recover $\mathbf{M} = \mathbf{G}_t^{-1}\,(\mathbf{G}_t\mathbf{M})$.

The use of packet headers incurs a transmission overhead relative to a fixed network (with the relevant global encoding vectors known *a priori* by all sinks); however, for certain practical transmission scenarios, where packet lengths can potentially be large, this overhead can be held to a small percentage of the total information transferred in a generation. For example, take $q = 2^8 = 256$ (field elements are bytes), a generation size $r = 50$, and packets of length $2^{11} = 2048$ bytes. The packet header must be 50 bytes, representing an overhead of just $50/2048 = 2.4\%$.

6.2. Subspace Transmission

Is this approach of sending packet headers fundamentally necessary in order to communicate over a noncoherent random linear network coding channel? Although the network will certainly impose structure on $\mathbf{G}_t$, one robust approach is to simply ignore such structure, treating $\mathbf{G}_t$ as random (but with full rank r with high probability, where r is the number of packets in the current generation). What property of $\mathbf{P}$ is preserved in $\mathbf{Y} = \mathbf{G}_t\mathbf{P}$, when $\mathbf{G}_t$ is a random matrix of rank r? Since $\mathbf{G}_t$ is random, all that is fixed by the product $\mathbf{G}_t\mathbf{P}$ is the row space of p. Indeed, as far as the sink node t is concerned, any of the possible generating sets for this space are equivalent.

This suggests an interesting approach for transmission of information over the noncoherent random linear network coding channel [24]. Rather than transmitting information by the choice $\mathbf{P}$, transmit information by the choice of the *vector space* spanned by the rows of $\mathbf{P}$. Even if $\mathbf{G}_t$ is not full rank, what is received is a *subspace* of the row space of **P**; such a dimensional "erasure" is potentially even correctable, if we can design a codebook (of vector spaces) in which such a lower-dimensional subspace is contained within just one codeword. It is even potentially possible to design a codebook to correct "errors," i.e., injection into the network (by an adversary or by channel noise) of packets *not* sent by the source.

This subspace approach to information transmission over the noncoherent random linear network coding channel has been developed in a series of papers [24, 25]. A family of error-and-erasure correcting codes with properties similar to Reed-Solomon codes have been constructed; such codes can be viewed as an adaptation of classically studied rank-metric codes [26] to the noncoherent network coding channel. Such codes can be used in schemes that provide robustness against errors and erasures, but also security against wiretappers [27] that may observe a (low rank) linear combination of the message vectors.

7. ON ALPHABETS AND NONLINEARITY

We have seen that the multicast capacity of a network can be achieved with linear network coding over a sufficiently large finite field. Indeed, a field size $q > |T|$ suffices.

Suppose that we wish to minimize the size of the alphabet used. As proved in [28], even deciding whether there exists a linear network code over F_q for a multicast network problem is NP-hard.

The following example, due to Riis and Ahlswede [29], gives a hint of why such problems about network coding alphabet size may be difficult. Consider the network shown in Fig. 1.6. Here we wish to communicate x and y (assumed to be elements from a finite set A) to five sinks, none of whom observe x and y directly, but rather who observe various mixtures of x, y, $x \oplus y$, and $x \boxplus y$. The functions $\oplus$ and $\boxplus$ are mappings from $A \times A$ to A, i.e., they are binary operators, to be specified.

What properties should the operators $\oplus$ and $\boxplus$ have? Clearly, since we wish to recover x knowing $x \oplus y$ and y, and we wish to recover y knowing $x \oplus y$ and x, the operation table for $\oplus$ must form a Latin square.[1] Similarly,

---

[1] Recall that a Latin square of order m is an $m \times m$ matrix such that each element from a set of size m appears exactly once in each row and column. The addition table for F_4, shown in Appendix A is an example of a 4×4 Latin square.

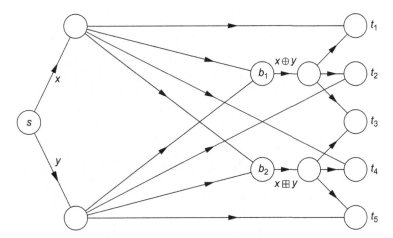

Figure 1.6 The sinks must reconstruct x and y from various combinations of x, y, $x \oplus y$, and $x \boxplus y$. However, $x \oplus y$ and $x \boxplus y$ are defined, this arithmetic must define an orthogonal Latin square, and so no solution exists over alphabets of size 2 or 6 elements.

the operation table for $\boxplus$ must also be a Latin square. In fact, since t_3 must be able to recover x and y from $x \oplus y$ and $x \boxplus y$, these operation tables must form so-called *orthogonal Latin squares*, i.e., the values of $x \oplus y$ and $x \boxplus y$ must uniquely determine x and y. Thus the question of whether the multicast problem of Fig. 1.6 can be solved over an alphabet A is equivalent to the question of whether there exist pairs of orthogonal Latin squares of order $|A|$.

The question of when pairs of orthogonal Latin squares exist dates back to Euler, but was not completely resolved until 1959 (see [29] for a discussion). It is now known that orthogonal Latin squares exist for all orders except $m = 2$ and $m = 6$. Thus the multicast problem of Fig. 1.6 has a solution if, and only if, the underlying alphabet does not have 2 or 6 elements. Note that the existence of a solution for an alphabet of size 3 does *not* imply existence of a solution for all larger alphabets.

The situation for non-multicast network communication problems is even more complicated, due mainly to the potential for "interference" among the various network flows. It is known that linear network coding does *not* suffice to solve such network information flow problems.

For example, consider the network shown in Fig. 1.7(a). There are three source nodes a, b, and c. Sink node demands are as indicated. Dougherty, Freiling and Zeger show in [30] that this network admits a scalar linear

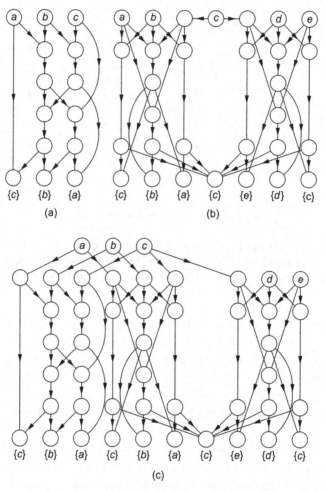

Figure 1.7 "Diabolical" network examples from [30]. For the sources and demands shown in the figure, the network (a) has a scalar linear solution only over a field of characteristic 2, (b) has a scalar linear solution only over a field of characteristic *not* 2, and (c), which combines the networks (1) and (2) has no scalar solution over any field, but admits a nonlinear solution over an alphabet of size 4.

solution only over a field of characteristic two. The network of Fig. 1.7(b), on the other hand, admits a scalar linear solution only over a field with a characteristic not equal to two. The network of Fig. 1.7(c), which combines these two networks, therefore admits no linear solution over *any* field (although a nonlinear solution over an alphabet of size 4 is given in [30]). This "diabolical" example rather convincingly demonstrates that linear network coding does *not* suffice to solve general network coding problems.

8. CONCLUSIONS

As we have seen in this chapter, the ideas that underly network coding are simple, yet potentially far-reaching. Allowing the nodes in a network to perform coding (not just routing), permits the multicast capacity of a network to be achieved with linear network coding, with potentially unbounded throughput advantage (in special cases) over routing. Unlike optimal routing solutions to a multicast problem, an optimal linear network coding solution can be found in polynomial time (or, with high probability of success, simply at random). In practice, the random linear network coding approach results in multicast protocols that are robust to changes in the underlying network topology. Despite its optimality for multicasting, it turns out that linear network coding does *not*, in general, suffice to achieve optimal throughputs for general flow problems, as illustrated by several carefully crafted "diabolical" examples.

The remainder of this book explores the use of network coding in a variety of application scenarios. I wish you happy reading!

ACKNOWLEDGMENTS

I am grateful for helpful comments made by Chen Feng, Ashish Khisti, Danilo Silva, and Alex Sprintson on an earlier version of this chapter. The proof of Lemma 3 in Appendix C is due to Joachim Rosenthal.

APPENDIX

A. Finite Fields

We will not attempt to be overly formal here. Algebraically, a *field* is a set of elements, together with two well-defined binary operations (addition, $+$, and multiplication, $\times$) that satisfy the familiar laws of arithmetic.

Addition in a field must be associative and commutative; a field must contain an additive identity (denoted 0); and each field element x must have an additive inverse (denoted $-x$). A more succinct statement is that the elements of a field must form an abelian group under addition.

Multiplication in a field must be associative and commutative; a field must contain a multiplicative identity (denoted 1); and each field element x (except 0) must have a multiplicative inverse (denoted x^{-1}). A more succinct statement is that the *nonzero* elements of a field must form an abelian group under multiplication.

Finally, multiplication must distribute over addition, so that $x \times (y + z) = (x \times y) + (x \times z)$ for all possible choices of field elements x, y, and z.

Subtraction and division in a field are defined via inverse elements, so that $x - y = x + (-y)$ and $x/y = x \times y^{-1}$, where the latter expression is defined only when $y \neq 0$. The multiplication symbol is often suppressed, i.e., if x and y are field elements, then xy denotes $x \times y$. For any positive integer a and field element x, we denote:

$$x^a = \underbrace{x \times x \times \cdots \times x}_{a \text{ times}}.$$

Familiar examples of (infinite) fields are the rational numbers, the real numbers, and the complex numbers, each with their usual addition and multiplication. The integers do *not* form a field, since, e.g., no nonzero integer (other than ± 1) has an integer multiplicative inverse. Of course a *finite field* is a field with a finite number of elements. If p is any prime, the set $\{0, 1, \ldots, p - 1\}$, closed under integer addition and multiplication modulo p, forms a field, denoted F_p.

For example, the smallest possible finite field, denoted F_2, contains just two elements: 0 and 1, and has the addition and multiplication tables (corresponding to integer addition and multiplication, modulo two) shown below. These correspond, respectively, to logical exclusive-OR (XOR) and logical AND. Note that $-1 = 1$ in F_2.

+	0	1		×	0	1
0	0	1		0	0	0
1	1	0		1	0	1

A number of facts follow immediately from the field axioms, and hence are true in any field F. For example, for all elements $x \in F$, $0 \times x = 0$ and $(-1) \times x = -x$. The product $x \times y$ of two elements is nonzero if, and only if, both x and y are nonzero; equivalently, the product $x \times y$ of two elements is zero if, and only if, at least one of x or y is zero.

Consider now the sequence $1, 1 + 1, 1 + 1 + 1, \ldots$, in a finite field F. Since F is finite, this sequence must eventually repeat, and it is easily shown that the first repeated element is 1, i.e., for some integer p, the elements $1, 2, 3, \ldots, p$ are distinct, but $p + 1 = 1$. The integer p is referred to as the *characteristic* of the finite field F. The characteristic of a finite field must be a prime number, since if it were the composite ab, then the expression:

$$\underbrace{(1 + 1 + \cdots + 1)}_{a \text{ times}} \underbrace{(1 + 1 + \cdots + 1)}_{b \text{ times}} = \underbrace{(1 + 1 + \cdots + 1)}_{ab \text{ times}}$$

would imply the existence of two nonzero elements whose product is zero, which is disallowed in a field. It can be shown that any two finite fields of the same size are isomorphic; i.e., apart from possible relabeling of the elements, there is only one finite field of any given size.

A *polynomial* over F in indeterminate x is an expression of the form $a(x) = \sum_{i \geq 0} a_i x^i$, with coefficients $a_i \in F$, only finitely many of which may be nonzero. The *degree* of a nonzero polynomial is the largest integer d such that $a_d \neq 0$. The set of all polynomials over F in indeterminate x is denoted as $F[x]$; this set is a ring under the familiar rules of polynomial addition and multiplication.

A polynomial $p(x) \in F[x]$ may, of course, be regarded as a function on F. Note, however, that distinct polynomials may not be distinct as functions; for example, both the zero polynomial and the polynomial $p(x) = x + x^2$ give rise to the same function on F_2. If $p(x) \in F[x]$ is a polynomial, and $p(a) = 0$ for some $a \in F$, then a is a *root* of $p(x)$. A polynomial of degree d can have no more than d roots.

If $p(x) = a(x)b(x)$ for some polynomials $a(x)$ and $b(x)$ of nonzero degree, then we say that $p(x)$ is reducible. If no such factorization exists, then $p(x)$ is irreducible. If F is finite, and m is any positive integer, it can be shown that an irreducible polynomial in $F[x]$ having degree m always exists.

When $a(x) \in F[x]$ is irreducible and of degree $m > 1$, the ring $F[x]/\langle a(x) \rangle$, which we may think of as the ring of polynomials under polynomial arithmetic reduced modulo $a(x)$, forms a field. The size of this field is $|F|^m$. Since irreducible polynomials of all possible degrees exist, starting from the finite field F_p, it is possible in this way to construct fields of size p^m, for any positive integer m. In fact, all finite fields can be constructed in this way: if F_q is a finite field of q elements, then $q = p^m$ for some prime p and some integer m, and F_q can be constructed as $F_p[x]/\langle a(x) \rangle$ for some irreducible polynomial $a(x)$.

To illustrate, it is easy to verify that $a(x) = x^2 + x + 1$ is irreducible in $F_2[x]$. Under polynomial arithmetic modulo $a(x)$, each element $p(x)$ of $F_2[x]$ falls into one of four residue classes: $[0]$; $[1]$; $[x]$; and $[1 + x]$, which is to say that the remainder of $p(x)$ upon division by $a(x)$ must be one of 0, 1, x, or $1 + x$. Under arithmetic modulo $x^2 + x + 1$, we obtain the addition and multiplication tables shown below. For example, we have $[x][x] = [1 + x]$, since $x^2 \bmod (x^2 + x + 1) = x + 1$. This is the field F_4, and larger fields can be obtained in a similar fashion.

$+$	$[0]$	$[1]$	$[x]$	$[1 + x]$
$[0]$	$[0]$	$[1]$	$[x]$	$[1 + x]$
$[1]$	$[1]$	$[0]$	$[1 + x]$	$[x]$
$[x]$	$[x]$	$[1 + x]$	$[0]$	$[1]$
$[1 + x]$	$[1 + x]$	$[x]$	$[1]$	$[0]$

$\times$	$[0]$	$[1]$	$[x]$	$[1 + x]$
$[0]$	$[0]$	$[0]$	$[0]$	$[0]$
$[1]$	$[0]$	$[1]$	$[x]$	$[1 + x]$
$[x]$	$[0]$	$[x]$	$[1 + x]$	$[1]$
$[1 + x]$	$[0]$	$[1 + x]$	$[1]$	$[x]$

Methods, such as Gauss-Jordan elimination, for solving systems of linear equations carry over to any field F. Matrices and determinants behave in the usual way; in particular, a square matrix with entries from F is invertible if and only if it has nonzero determinant.

B. Zeros and Nonzeros of Polynomials

When dealing with multivariate polynomials in $F[x_1, \ldots, x_n]$, the notion of "degree" becomes less straightforward than in the univariate case. A monomial $x_1^{d_1} x_2^{d_2} \cdots x_n^{d_n}$, where each d_i is a non-negative integer, can be regarded as having x_1-degree equal to d_1 (when regarded as a polynomial in x_1), x_2-degree d_2, etc., or *total degree* $d_1 + d_2 + \cdots + d_n$. Likewise, the x_i-degree of a polynomial $p(x_1, \ldots, x_n) \in F[x_1, \ldots, x_n]$ is the largest among the x_i-degrees of the monomials that appear as nonzero terms in p, and the total degree of p is the largest among the total degrees of the monomials that appear as nonzero terms in p. If the x_i-degree of p is never larger, for any i, than a fixed integer d, then we say that p has degree bounded by d. Necessarily, p then has total degree bounded by nd.

The following well-known lemma (see, e.g., [31, Lemma 2.1]) is equivalent to the Sparse Zeros Lemma.

Lemma 2. *Let $f \in F[x_1, \ldots, x_n]$ have degree bounded by d and let $A \subseteq F$ be a set of $d + 1$ distinct elements of F. If $f(a_1, \ldots, a_n) = 0$ for all n-tuples $(a_1, \ldots, a_n)$ in A^n, then $f \equiv 0$, i.e., f is identically the zero polynomial.*

Proof. We proceed by induction on n, the number of variables. Recall that a nonzero univariate polynomial of degree at most d over F can have at most d zeros; this establishes the truth of the proposition for $n = 1$.

Suppose, for some $n \geq 1$, that the proposition is true for all polynomials of degree bounded by d in $F[x_1, \ldots, x_n]$. Let $f \in F[x_1, \ldots, x_{n+1}]$ have degree bounded by d, written as:

$$f(x_1, \ldots, x_{n+1}) = \sum_{i=0}^{d} f_i(x_1, \ldots, x_n) x_{n+1}^i,$$

where $f_i \in F[x_1, \ldots, x_n]$ are polynomials of degree bounded by d. Suppose that $f(a_1, \ldots, a_{n+1}) = 0$ for all $(a_1, \ldots, a_{n+1}) \in A^{n+1}$. Then, in particular, each polynomial $f(a_1, \ldots, a_n, x_{n+1})$ in x_{n+1} (fixing $(a_1, \ldots, a_n)$) has at least $d + 1$ zeros, and hence must identically be the zero polynomial. This implies that $f_i(a_1, \ldots, a_n) = 0$ for all $(a_1, \ldots, a_n) \in A^n$. By assumption, each such f_i must be identically the zero polynomial, which implies that f is identically the zero polynomial.

The proposition now follows by induction. ∎

The Sparse Zeros Lemma used in Section 3.3 is simply the contrapositive statement with $A = F$.

The following theorem is also useful in the analysis of random linear network coding.

Theorem 2 (Schwartz–Zippel). *Let $f \in F[x_1, \ldots, x_n]$ be nonzero with total degree d. If values $r_1, \ldots, r_n$ are chosen uniformly at random from a set $A \subseteq F$, then the probability that $f(r_1, \ldots, r_n) = 0$ is at most $d/|A|$.*

Proof: (by induction on n) In case $n = 1$, f is univariate with at most d roots. Even if A happens to contain all d of them, the probability that r_1 is chosen equal to a root is at most $d/|A|$.

Suppose that the induction hypothesis holds for polynomials with at most $n - 1$ variables, where $n > 1$. Express $f(x_1, \ldots, x_n)$ as a polynomial in x_1, i.e., write:

$$f(x_1, \ldots, x_n) = \sum_{i=0}^{k} x_1^i f_i(x_2, \ldots, x_n)$$

where $k \leq d$ is the x_1-degree of f. For this choice of k, $f_k(x_2, \ldots, x_n)$ cannot be identically zero. The total degree of f_k is at most $d - k$, so the probability that $f_k(r_2, \ldots, r_n) = 0$ is at most $(d - k)/|A|$.

Now if $f_k(r_2, \ldots, r_n)$ is not equal to zero, let:

$$g(x_1) = f(x_1, r_2, \ldots, r_n) = \sum_{i=0}^{k} x_1^i f_i(r_2, \ldots, r_n).$$

This polynomial has degree k and is not identically zero, hence the probability that $g(r_1)$ evaluates to zero is at most $k/|A|$.

Let B be the event that $f(r_1, \ldots, r_n) = 0$ and let C be the event that $f_k(r_2, \ldots, r_n) = 0$. We have $P[C] \leq (d - k)/|A|$ and $P[B \mid \overline{C}] \leq k/|A|$. Thus:

$$
\begin{aligned}
P[B] &= P[B \mid C]P[C] + P[B \mid \overline{C}]P[\overline{C}] \\
&\leq P[C] + P[B \mid \overline{C}] \\
&\leq (d - k)/|A| + k/|A| \\
&= d/|A|.
\end{aligned}
$$

∎

Ho, *et al.* [16] give the following refinement of the Schwartz-Zippel Theorem.

Theorem 3. *Let* $f \in F[x_1, \ldots, x_n]$ *be a polynomial not identically zero with degree bounded by d and with total degree bounded by $d\eta$. If values $r_1, \ldots, r_n$ are chosen uniformly at random from a set $A \subseteq F$, with $|A| \geq d$, then the probability that $f(r_1, \ldots, r_n) = 0$ is at most $1 - (1 - d/|A|)^\eta$.*

C. The Degree of $\det(\mathbf{G}_t)$

We start with the following lemma.

Lemma 3. *Let* $\mathbf{A}$ *and* $\mathbf{B}$ *be* $n \times n$ *matrices with entries from a field, and let x be an indeterminate. Then* $\det(\mathbf{A}x + \mathbf{B})$ *is a polynomial in x of degree at most* rank$(\mathbf{A})$.

Proof. The statement is trivially true if $\mathbf{A}$ has rank zero.

Suppose that $\mathbf{A}$ has rank one. Let $\mathbf{A} = [\mathbf{a}_1, \ldots, \mathbf{a}_n]$, and let $\mathbf{B} = [\mathbf{b}_1, \ldots, \mathbf{b}_n]$, where $\mathbf{a}_i$ and $\mathbf{b}_i$ denote the ith column of $\mathbf{A}$ and $\mathbf{B}$, respectively. Since the determinant of a matrix is multilinear in the rows or columns of that matrix, we have:

$$\det(\mathbf{A}x + \mathbf{B}) = \det(\mathbf{a}_1 x + \mathbf{b}_1, \ldots, \mathbf{a}_n x + \mathbf{b}_n)$$
$$= \det(\mathbf{b}_1, \ldots, \mathbf{b}_n) + x \det(\mathbf{a}_1, \mathbf{b}_2, \ldots, \mathbf{b}_n)$$
$$+ x \det(\mathbf{b}_1, \mathbf{a}_2, \mathbf{b}_3, \ldots, \mathbf{b}_n) + \cdots$$
$$+ x \det(\mathbf{b}_1, \ldots, \mathbf{b}_{n-1}, \mathbf{a}_n) + \cdots$$

Since rank$(\mathbf{A}) = 1$, terms in this expansion involving determinants of matrices that include two or more columns from $\mathbf{A}$ are zero. Thus the given determinant has x-degree at most one.

In a similar manner one can see that if $\mathbf{A}$ has rank $r > 1$, then all terms in the multilinear determinant expansion involving matrices with more than r columns from $\mathbf{A}$ will be zero, which means that the resulting determinant has x-degree at most r. ∎

Theorem 4. *Let* $\mathbf{G}_t(\alpha_1, \ldots, \alpha_\eta)$ *be a global transfer matrix for an acyclic network, as defined in Section 3.2. Then* $\det(\mathbf{G}_t)$ *is a polynomial of degree bounded by one.*

Proof. The (i,j)th entry of $\mathbf{G}_t$ is the sum over all paths in G' from input node i to output node j (at sink node t) of the product of branch gains along each path. Let e denote the (one) edge in the line graph G' having branch gain α_i. By partitioning paths into two disjoint sets (those that include e and those that do not), we may write:

$$\mathbf{G}_t = \alpha_i \mathbf{x}\mathbf{y}^T + \mathbf{H}$$

where $\mathbf{H}$ is the $r \times r$ matrix obtained from $\mathbf{G}_t$ by setting $\alpha_i = 0$, $\mathbf{x}$ is the $r \times 1$ vector of gains from input nodes to e and $\mathbf{y}$ is the $r \times 1$ vector of gains from e to the output nodes. Note that $\mathbf{H}$ is constant with respect to α_i. The influence of α_i is a rank-1 update to $\mathbf{H}$, and so the theorem follows from Lemma 3 (with $x = \alpha_i$, $\mathbf{A} = \mathbf{x}\mathbf{y}^T$ and $\mathbf{B} = \mathbf{H}$).   ∎

REFERENCES

[1] R. Ahlswede, N. Cai, S.-Y. R. Li, and R. W. Yeung, "Network information flow," *IEEE Trans. on Inform. Theory*, vol. 46, pp. 1204–1216, July 2000.

[2] R. W. Yeung, S.-Y. R. Li, N. Cai, and Z. Zhang, "Network coding theory," *Foundations and Trends in Communications and Information Theory*, vol. 2, no. 4,5, pp. 241–381, 2006.

[3] C. Fragouli and E. Soljanin, "Network coding fundamentals," *Foundations and Trends in Networking*, vol. 2, no. 1, pp. 1–133, 2007.

[4] C. Fragouli and E. Soljanin, "Network coding applications," *Foundations and Trends in Networking*, vol. 2, no. 2, pp. 135–269, 2007.

[5] T. Ho and D. S. Lun, *Network Coding: An Introduction*. Cambridge, UK: Cambridge University Press, 2008.

[6] T. M. Cover and J. A. Thomas, *Elements of Information Theory*. Hoboken, NJ: John Wiley and Sons, second ed., 2006.

[7] K. Menger, "Zur allgemeinen Kurventheorie," *Fundamenta Mathematicae*, vol. 10, pp. 96–115, 1927.

[8] L. R. Ford, Jr. and D. R. Fulkerson, "Maximal flow through a network," *Canadian J. Mathematics*, vol. 8, pp. 399–404, 1956.

[9] P. Elias, A. Feinstein, and C. E. Shannon, "A note on the maximum flow through a network," *IRE Trans. on Inform. Theory*, vol. 2, no. 4, pp. 117–119, 1956.

[10] S.-Y. R. Li, R. W. Yeung, and N. Cai, "Linear network coding," *IEEE Trans. on Inform. Theory*, vol. 49, pp. 371–381, Feb. 2003.

[11] R. Koetter and M. Médard, "An algebraic approach to network coding," *IEEE/ACM Trans. on Networking*, vol. 11, pp. 782–795, Oct. 2003.

[12] D. B. West, *Introduction to Graph Theory*. Upper Saddle River, NJ: Prentice Hall, second ed., 2001.

[13] S. Jaggi, P. Sanders, P. A. Chou, M. Effros, S. Egner, K. Jain, and L. M. G. M. Tolhuizen, "Polynomial time algorithms for multicast network code construction," *IEEE Trans. on Inform. Theory*, vol. 51, pp. 1973–1982, June 2005.

[14] N. J. A. Harvey, D. R. Karger, and K. Murota, "Deterministic network coding by matrix completion," in *Proc. 16th Annual ACM-SIAM Symp. on Discrete Algorithms (SODA)*, (Vancouver, BC), pp. 489–498, Jan. 23–25, 2005.

[15] Á. I. Barbero and Ø. Ytrehus, "Heuristic algorithms for small field multicast encoding," in *Proc. 2006 IEEE Inform. Theory Workshop*, (Chengdu, China), pp. 428–432, Oct. 22–26, 2006.

[16] T. Ho, M. Médard, R. Koetter, D. R. Karger, M. Effros, J. Shi, and B. Leong, "A random linear network coding approach to multicast," *IEEE Trans. on Inform. Theory*, vol. 52, pp. 4413–4430, Oct. 2006.

[17] R. M. Roth, *Introduction to Coding Theory*. Cambridge, UK: Cambridge University Press, 2006.

[18] K. Jain, M. Mahdian, and M. R. Salavatipour, "Packing Steiner trees," in *Proc. 14th Annual ACM-SIAM Symp. on Discrete Algorithms (SODA)*, (Baltimore, MD), pp. 266–274, Jan. 12–14, 2003.

[19] A. Agarwal and M. Charikar, "On the advantage of network coding for improving network throughput," in *Proc. 2004 IEEE Inform. Theory Workshop*, (San Antonio, TX), pp. 247–249, Oct. 24–29, 2004.

[20] E. Halperin, G. Kortsarz, R. Krauthgamer, A. Srinivasan, and N. Wang, "Integrality ratio for group Steiner trees and directed Steiner trees," in *Proc. 14th Annual ACM-SIAM Symp. on Discrete Algorithms (SODA)*, (Baltimore, MD), pp. 275–284, Jan. 12–14, 2003.

[21] L. Zosin and S. Khuller, "On directed Steiner trees," in *Proc. 13th Annual ACM-SIAM Symp. on Discrete Algorithms (SODA)*, (San Francisco, CA), pp. 59–63, Jan. 6–8, 2002.

[22] C. Chekuri, C. Fragouli, and E. Soljanin, "On average throughput and alphabet size in network coding," *IEEE Trans. on Inform. Theory*, vol. 52, pp. 2410–2424, June 2006.

[23] P. A. Chou, Y. Wu, and K. Jain, "Practical network coding," in *Proc. 2003 Allerton Conf. on Commun., Control, and Computing*, (Monticello, IL), pp. 40–49, Oct. 1–3, 2003.

[24] R. Kötter and F. R. Kschischang, "Coding for errors and erasures in random network coding," *IEEE Trans. on Inform. Theory*, vol. 54, pp. 3579–3591, Aug. 2008.

[25] D. Silva, F. R. Kschischang, and R. Koetter, "A rank-metric approach to error control in random network coding," *IEEE Trans. on Inform. Theory*, vol. 54, pp. 3951–3967, Sept. 2008.

[26] E. M. Gabidulin, "Theory of codes with maximum rank distance," *Probl. Peredachi Informatsii*, vol. 21, no. 1, pp. 3–16, 1985.

[27] D. Silva and F. R. Kschischang, "Universal secure network coding via rank-metric codes," *IEEE Trans. on Inform. Theory*, vol. 57, no. 2, pp. 1124–1135, Feb. 2011.

[28] A. Rasala Lehman and E. Lehman, "Complexity classification of network information flow problems," in *Proc. 15th Annual ACM-SIAM Symp. on Discrete Algorithms (SODA)*, (New Orleans, LA), pp. 142–150, Jan. 11–13, 2004.

[29] S. Riis and R. Ahlswede, "Problems in network coding and error correcting codes," in *Proc. First Workshop on Network Coding, Theory, and Applications*, (Riva del Garda, Italy), Apr. 7, 2005.

[30] R. Dougherty, C. Freiling, and K. Zeger, "Insufficiency of linear coding in network information flow," *IEEE Trans. on Inform. Theory*, vol. 51, pp. 2745–2759, Aug. 2005.

[31] N. Alon, "Combinatorial nullstellensatz," *Combinatorics, Probability and Computing*, vol. 8, pp. 7–29, 1999.

CHAPTER 2

Harnessing Network Coding in Wireless Systems

Dina Katabi[1], Sachin Katti[2], and Hariharan Rahul[1]
[1]Computer Science & Artificial Intelligence Lab, MIT, MA, USA;
[2]Department of Electrical Engineering, Stanford University, CA, USA

Contents

Abstract

Wireless networks suffer from a variety of unique problems including low through-put, dead spots, and interference. However, their characteristics, such as the broadcast nature of the medium, spatial diversity, and significant data redundancy, provide opportunities for new design principles to address these problems. In this chapter, we describe advances in employing network coding to improve the throughput and reliability of wireless networks.

Keywords: Wireless networks, mesh networks, empirical demonstration, system architecture, network protocols.

Network Coding. DOI: 10.1016/B978-0-12-380918-6.00002-0
Copyright © 2012 Elsevier Inc. All rights reserved.

1. INTRODUCTION

Wireless networks have been designed using the wired network as the blueprint. The design abstracts the wireless channel as a point-to-point link, and grafts wired network protocols onto the wireless environment. For example, routing uses shortest path protocols, routers forward packets but do not modify the data, and reliability relies on retransmissions. The design has worked well for wired networks, but less so for the unreliable and unpredictable wireless medium.

The wireless medium is fundamentally different. While wired networks have reliable and predictable links, wireless links have high bit error rate, and their characteristics could vary over short time scales. Further, wired links are unicast links, but the majority of wireless links (with omni-directional antennas) are broadcast links. Transmissions in a wired network do not intefere with each other, whereas interference is a common case for the wireless medium. Wired nodes are usually static, while wireless was built to support mobility and portability. The wired network design conflicts with the characteristics of the wireless medium. As a result, current wireless networks suffer low throughput, dead spots, and inadequate mobility support.

The characteristics of wireless networks might all seem disadvantageous at first sight, but a newer perspective reveals that some of them can be used to our advantage, albeit with a fresh design. The broadcast nature of wireless provides an opportunity to deal with unreliability; when a node broadcasts a packet, it is likely that at least one nearby node receives it, which can then function as the next-hop and forward the packet. This is in stark contrast to the present wireless design, where there is a single designated next-hop, and when it does not receive the packet, the previous hop has to retransmit it. The property is called *spatial diversity*. A naive approach to exploiting spatial diversity would make any node that overhears a packet forward it toward its destination. This can lead to multiple nodes forwarding the same packet unnecessarily and wasting network bandwidth. Ideally, one would like to have a distributed coding scheme, where the relaying nodes code the data in the forwarded packets so that all transmitted information is useful.

Additionally, wireless networks exhibit significant data redundancy, i.e., there is a large overlap in the information available to the nodes. First, as a packet travels multiple hops, its content becomes known to many nodes.

Further, wireless broadcast amplifies this redundancy because at each hop it delivers the same packet to multiple nodes within the transmitter's radio range. Thus, as a node considers a packet for forwarding, it can leverage that multiple nearby nodes already know the packet's content, i.e., they have side information. Thus, one may want to code the transmission in a manner that leverages this side information to perform in-network compression of the data.

Can a network coding design of wireless networks exploit their intrinsic characteristics, such as spatial diversity and data redundancy, to improve their throughput and reliability? In this chapter, we survey a few recent advances in employing network coding to increase the throughput and reliability of wireless networks. We focus on experimental wireless systems. These systems complement the theoretical work discussed in other chapters. They build on the concept of network coding to deliver new algorithms that yield themselves to practical implementations, and system foundations that integrate network coding into the existing network stack. The practicality of these algorithms is demonstrated via prototype implementations and testbed evaluations.

The emphasis of this chapter will be on the first wireless systems that demonstrated empirically the benefits of various types of network coding, including: inter-session coding, intra-session coding, and analog network coding.

2. NETWORK CODING BACKGROUND: THE PRACTITIONER'S PERSPECTIVE

Traditional computer networks have a store-and-forward design, i.e., routers buffer received packets and forward them unmodified. Network coding changes this design in a simple yet basic way. It enables the routers to code or mix packets before forwarding, even when the packets are from otherwise unrelated communication streams. Both theoretical analysis and practical systems show that network coding can produce dramatic improvements in the throughput and the reliability of wireless networks [5, 14, 17, 21].

One may identify three approaches to network coding: intra-session coding, inter-session coding, and analog network coding.

(a) **Intra-Session Network Coding:** In this approach, routers mix packets addressed to the same destination. Routers need not decode the packets; the destination eventually decodes them once it receives a sufficient number of coded packets [2]. Early work on network coding has been theoretical and focused on intra-session network coding. Two of these theoretical results have important practical implications. First, the combination of [12, 18, 19] shows that, for multicast traffic, linear codes (i.e., having the routers create linear combinations of the packets they receive, $\sum c_i p_i$, where p_i is a packet and c_i is some coefficient) achieve the maximum capacity bounds, and coding and decoding can be done in polynomial time. Second, Ho *et al.* show that the above is true even when the routers pick random coefficients [10]. This enables distributed network coding, where routers do not need to consult each other on the choice of codes.

Researchers have applied intra-session network coding to many areas including energy [8, 28], secrecy [4, 11], content distribution [6], and distributed storage [13]. In this chapter, however, we focus on its application to wireless networks [5, 21, 26]. In particular, we discuss two protocols: MORE [5], and MIXIT [15]. The former delivers the first implementation of intra-session network coding in a wireless network. The latter is a cross-layer approach to intra-session network coding, that allows the network to take advantage of all the correctly received bytes, including correct bytes in corrupted packets.

(b) **Inter-Session Network Coding:** In this approach, routers mix or code packets that belong to different streams of communications that intersect at an intermediate point in the network. Consider the example in Fig. 2.1, where Alice and Bob want to exchange a couple of packets. The radio range does not permit them to communicate directly, and thus they need

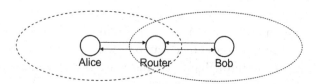

Figure 2.1 A simple example of how network coding can improve throughput. It allows Alice and Bob to exchange a pair of packets through *R* using 3 transmissions instead of 4.

to go through a router. In the current design, Alice sends her packet to the router, which forwards it to Bob, and Bob sends his packet to the router, which forwards it to Alice. Yet, as noted in [16, 17], Alice and Bob could transmit their respective packets to the router, which XORs the two and broadcasts the resulting coded packet. Alice recovers Bob's packet by XOR-ing again with her own, and Bob recovers Alice's packet in the same way. The process exploits the existing redundancy in the network to compress the information, delivering two packets in a single transmission, and improving the throughput. This example was first introduced in [27]. It was generalized to various network topologies and traffic patterns in [17]. The resulting protocol, COPE, delivered the first implementation of network coding, which demonstrated practical and significant throughput gains.

In contrast to its intra-session counterpart, inter-session network coding was introduced as a practical protocol and system architecture in [17]. Its practical success has motivated extensive theoretical work that characterizes the throughput bounds of this approach [20, 25].

(c) Analog Network Coding (ANC): Analog Network Coding (ANC) extends network coding to operate over analog signals rather than packets [14]. If two packets are transmitted simultaneously, the resulting interfered signal is a linear combination of the two native signals. Thus, instead of having the routers mix or code packets, analog network coding mixes packets' content directly over the channel. This reduces the number of transmissions since it allows multiple nodes to transmit simultaneously and still have their packets delivering useful information.

Analog network coding is also referred to as physical-layer network coding, which first appeared in [24]. The design proposed in [24], however, does not support a practical implementation. It assumes symbol-level synchronization, carrier-frequency synchronization, and carrier-phase synchronization, which are unlikely in realistic settings. In [14], Katti *et al.* have developed a decoding algorithm that works despite channel distortion and lack of synchronization between transmitters. Based on this algorithm, they demonstrated the first implementation of analog network coding.

It is worth noting that analog network coding can be applied to packets within a session and across sessions, as will be discussed later.

Table 2.1 compares intra- and inter-session network coding.

Table 2.1 Comparison between different approaches to network coding

Intra-session Network Coding	Inter-session Network Coding	Analog Network Coding
Codes packets from the same session	Codes packets from different sessions	Code signals by transmitting them concurrently
Decoding can be done at destination(s)	Routers typically decode and re-encode the packets	Routers need not decode, they amplify and forward mixed signals
Typically used for multicast applications	Typically used for unicast applications	Used for both unicast and multicast
Suitable for lossy links and unpredictable topologies	Suitable for low-loss static topologies	Suitable for high SNR channels
Improves reliability	Improves throughput	Improves throughput

3. APPLICATIONS OF NETWORK CODING IN WIRELESS NETWORKS

The following sections describe the potential benefits of building future wireless networks around network coding. We weave the argument by considering different axes of a wireless network—throughput, reliability, and dealing with interference. Each of these axes harnesses a different approach to network coding.

3.1. COPE: Network Coding for Increased Throughput

The throughput of today's wireless networks leaves a lot to be desired; we summarize here COPE [17], a wireless architecture designed around network coding that improves wireless throughput. COPE is the first general algorithm that codes packets across multiple connections while making no assumptions about topology, traffic demands, or scheduling. It also presents the first design that integrates network coding in the current network stack, and the first implementation and evaluation of network coding in a testbed [16, 17].

The intuition underlying COPE is that network coding increases wireless throughput because coding allows the routers to compress the transmitted information given what is known at various nodes, as in the

Alice-Bob scenario in Fig. 2.1. By matching what each neighbor has with what another neighbor wants, a router can deliver multiple packets to different neighbors in a single transmission. This style of coding is called *inter-session network coding* because the coding is done over packets that differ in their next-hop, and thus are from different sessions.

COPE can further extend the idea in Fig. 2.1 beyond two nodes communicating on reverse paths. To do so, COPE exploits the shared nature of the wireless medium, which, for free, broadcasts each packet in a small neighborhood around its path. This creates an environment conducive for coding because nodes in each area have a large and partially overlapping reservoir of packets they can use to decode. For example, Fig. 2.2 shows how two connections traversing different paths can be encoded together. Assume sources S1 and S2 are sending a stream of packets to destinations D_1 and D_2 respectively. Since wireless is a broadcast medium, by snooping, node D_1 hears S2's packets, and D_2 hears S1's packets traversing their neighborhood. Router R can exploit these overheard packets to XOR S1's and S2's packets and broadcast an XOR-ed version, which is useful to both destinations. The "X" topology in Fig. 2.2 is an extension of the Alice-Bob topology. It is an important extension because in a real wireless network, there might be only a small number of flows traversing the reverse path of each other *a la* Alice-Bob, but one would expect many flows to intersect at a router, and thus they can be coded together.

For more general topologies, COPE leads to larger bandwidth savings than are apparent from the above example. It can XOR more than a pair of packets and produce a multifold increase in the throughput.

3.1.1 The Protocol
COPE is a new forwarding architecture for wireless mesh networks. It inserts a coding layer between the IP and MAC layers, which detects coding

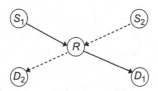

Figure 2.2 "X" Topology: Two flows intersecting at *R*.

opportunities and exploits them to forward multiple packets in a single transmission. COPE incorporates the following three main techniques:

(a) **Opportunistic Listening:** COPE requires nodes to listen and store packets that can be potentially used later for decoding coded packets. There are two sources for such packets:

- *Packets the node itself has broadcast.* For example, in the Alice–Bob example in Fig. 2.1, Alice and Bob keep a copy of the packets they transmitted to the router. They use these packets later for decoding the coded packet from the router.
- *Packets the node has overheard.* Since wireless is a broadcast medium, it creates many opportunities for nodes to overhear transmitted packets. In today's design, nodes throw away packets that they overhear but which are not meant for them. In COPE, however, nodes are kept in promiscuous mode, and they snoop on all communications over the wireless medium and store the overheard packets for a limited period T. The value of T should be larger than the maximum one-way latency in the network (the default is $T = 0.5s$).

(b) **Opportunistic Coding:** The key question is what packets to code together to maximize throughput. A node may have multiple options but it should aim to maximize the number of packets delivered in a single transmission, while ensuring that each intended next-hop has enough information to decode its intended packet.

The above is best illustrated with an example. In Fig. 2.3(a), node B has 4 packets in its output queue p_1, p_2, p_3, and p_4. Its neighbors have overheard some of these packets. The table in Fig. 2.3(b) shows the next-hop of each packet in B's queue. When the MAC permits B to transmit, B takes packet p_1 from the head of the queue. Assuming that B knows which packets each neighbor has, it has a few coding options, as shown in Fig. 2.3(c).

1. It could send $p_1 \oplus p_2$. Since node C has p_1 in store, it could XOR p_1 with $p_1 \oplus p_2$ to obtain the original packet sent to it, i.e., p_2. However, node A does not have p_2, and so cannot decode the XOR-ed packet. Thus, sending $p_1 \oplus p_2$ would be a bad coding decision for B, because only one neighbor can benefit from this transmission.
2. The second option in Fig. 2.3(c) shows a better coding decision for B. Sending $p_1 \oplus p_3$ would allow both neighbors C and A to decode and obtain their intended packets from a single transmission.

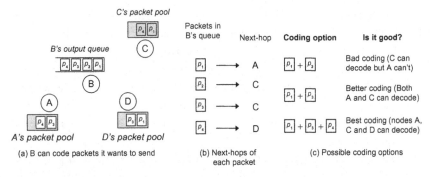

Figure 2.3 Example of COPE Coding; Node B has 4 packets in its queue, whose next-hops are listed in (b). Each neighbor of B has stored some packets as depicted in (a). Node B can make a number of coding decisions (as shown in (c)), but should select the last one because it maximizes the number of packets delivered in a single transmission.

3. However, the best coding decision for B would be to send $p_1 \oplus p_3 \oplus p_4$, which would allow all three neighbors to receive their respective packets from one transmission. For example, node A can XOR the received coded packet with p_3 and p_4 from its store to recover p_1, the packet meant for it. Neighbors C and D similarly can recover their intended packets.

The above example emphasizes an important coding issue. Packets from multiple unicast sessions may get encoded together at some intermediate hop. But their paths may diverge at the next-hop, at which point they need to be decoded. If not, unneeded data will be forwarded to areas where there is no interested receiver, wasting precious capacity. The coding algorithm should ensure that all next-hops of an encoded packet can decode their corresponding native packets. This can be achieved using the following simple rule:

To transmit n packets, $p_1, \ldots, p_n$, to n next-hops, $r_1, \ldots, r_n$, a node can XOR the n packets together only if each next-hop r_i has all $n - 1$ packets p_j for $j \neq i$.

This rule ensures that each next-hop can decode the XOR-ed version to extract its native packet. Whenever a node has a chance to transmit a packet, it chooses the largest n that satisfies the above rule to maximize the benefit of coding.

(c) **Learning Neighbor State:** To implement the coding algorithm described above, nodes in COPE need to know what packets their neighbors store. COPE achieves this through two mechanisms:

- *Reception Reports:* Nodes broadcast reception reports to tell their neighbors which packets they have in store. Reception reports are sent by annotating the data packets the node transmits. A node that has no data packets to transmit periodically sends the reception reports in special control packets.

- *Guessing:* If a node has not received a reception report for a particular packet in its forwarding queue, it guesses whether a neighbor has it. To guess intelligently, COPE leverages the routing computation. Wireless routing protocols compute the delivery probability between every pair of nodes and use it to identify good paths. For example, the ETX metric [7] periodically computes the delivery probabilities and assigns to each link a weight equal to *1/(delivery probability)*. These weights are broadcast to all nodes in the network and used by a link-state routing protocol to compute the shortest paths. We leverage these probabilities for guessing. In the absence of deterministic information, COPE estimates the probability that a particular neighbor has a packet as the delivery probability of the link between the packet's previous hop and the neighbor.

Guessing allows COPE to operate efficiently when reception reports get lost or are late. For example, at times of severe congestion, reception reports may get lost in collisions, while at times of light traffic, they may not arrive before a node has to make a coding decision on a particular packet. COPE is not handcuffed in such situations, since the guessing method gives a fairly accurate picture of neighbor state. Occasionally, a node may make an incorrect guess, which causes the coded packet to be undecodable at some next-hop. In this case, the relevant original packet is retransmitted, potentially encoded with a new set of original packets.

3.1.2 Performance Results

Theoretically, COPE reduces the number of transmissions by a factor of two [17], and thus should double the throughput. Katti *et al.* [17] have shown, however, that in practice the throughput gain is much larger. Figure 2.4 plots the aggregate throughput as a function of the traffic demands, for a 20-node mesh testbed. The figure shows that COPE brings

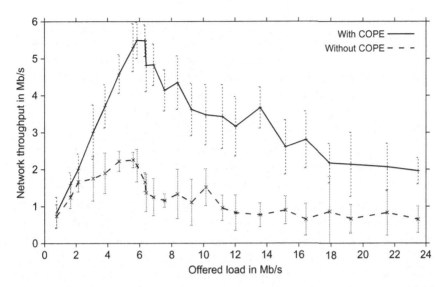

Figure 2.4 COPE, an inter-session network coding protocol, provides a several-fold increase in the throughput of wireless *ad hoc* networks. Results are for UDP sessions with randomly picked source-destination pairs, Poisson arrivals, and heavy-tail size distribution.

a *3–4x* increase in the throughput, when compared to current 802.11. Experimental throughput gain exceeds the theoretical gain because COPE alleviates hot-spots in the network. Specifically, in an *ad hoc* network, most paths cross at the center. As a result, nodes in the center of the network experience congestion, build queues, and drop packets. These dropped packets have consumed bandwidth to reach the center of the network. Dropping them in the center wastes network resources and significantly reduces the overall throughput. In contrast, with coding, congested nodes in the center of the network have the opportunity to send multiple packets in a single transmission, allowing them to drain their queues faster and avoid dropping packets.

3.2. MORE: Network Coding for Increased Reliability

The primary means of ensuring reliability in the present wireless architecture is retransmission of lost packets. This works well in wired networks where the bit error rate is very low, but is inefficient over error prone wireless channels.

This section describes how network coding can provide a more efficient approach to reliability. In contrast to Section 3.1., this section employs *intra-flow network coding*, i.e., routers mix packets heading to the same destination.

Our description focuses on MORE [5], the first implemented intra-session network coding protocol. MORE is an opportunistic routing protocol. Specifically, outdoor wireless mesh networks tend to have poor link quality caused by urban structures and the many interferers, including local WLANs [1]. Traditional routing chooses the next-hop before transmitting a packet; but, when link quality is poor, the probability that the chosen next-hop receives the packet is low. In contrast, opportunistic routing allows *any* node that overhears the transmission and is closer to the destination to participate in forwarding the packet [3]. As a result, opportunistic routing protocols are significantly more resilient to packet loss. Opportunistic routing, however, introduces a difficult challenge. Multiple nodes may hear a packet broadcast and unnecessarily forward the same packet. As a result, prior work on opportunistic routing requires global coordination to decide which node is responsible for forwarding a particular packet. In contrast, MORE employs network coding to solve this problem in a scalable manner without node coordination. Before explaining how MORE works we explain its design using a few examples.

3.2.1 Example 1: Dead Spots

Network coding also helps in dealing with dead spots. Consider the scenario in Fig. 2.5(a), where Alice would like to transfer a file to Bob. Bob is not within Alice's radio range, and thus Alice needs the help of an intermediate node. There are five nearby wireless nodes which could relay Alice's packets to Bob. Unfortunately, Alice is in a dead spot with 80% loss rate to every nearby wireless node. In today's 802.11 networks, Alice will pick the best path to Bob for her transfer. But since all paths are lossy, each packet will have to be transmitted six times (five times from Alice to the relay and once from the relay to Bob.)

A better approach would make use of spatial diversity to improve Alice's throughput. Alice broadcasts her packets, and any relay that hears a packet can forward it to Bob. In this case, the probability of delivering a packet to the relay increases from 20% to $(1 - 0.8^5) \times 100 \approx 67.2\%$. Thus, on average a packet is transmitted 2.5 times (1.5 times from Alice to the relays and once from the receiving relay to Bob.) This increases

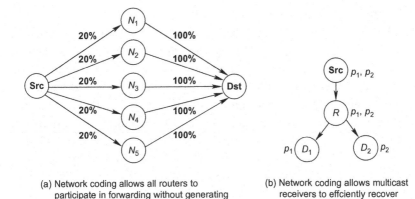

(a) Network coding allows all routers to
participate in forwarding without generating
duplicate transmissions

(b) Network coding allows multicast
receivers to effciently recover
lost packets

Figure 2.5 Example scenarios where network coding provides reliability benefits.

Alice's throughput by 2.4x. But using diversity without coding can create a different problem. Multiple relay nodes can hear the same packet and thus attempt to transmit it to Bob. This creates spurious transmissions and wastes the wireless bandwidth.

The combination of spatial diversity and network coding solves the problem. Alice broadcasts her packets. Any relay can participate in forwarding Alice's packets to Bob. To do so, the relay creates random linear combinations of the packets it has received from Alice so far. Specifically, a relay transmits to Bob packets of the form $p'_i = \sum c_{ij}p_j$, where p_j is one of Alice's packets, and c_{ij}'s are random coefficients. If the file contains n packets, then Bob can decode the whole file after receiving any n coded packets, as follows:[1]

$$\begin{pmatrix} p_1 \\ \vdots \\ p_n \end{pmatrix} = \begin{pmatrix} c_{11} & \cdots & c_{1n} \\ \vdots & \ddots & \\ c_{n1} & \cdots & c_{nn} \end{pmatrix}^{-1} \begin{pmatrix} p'_1 \\ \vdots \\ p'_n \end{pmatrix} \tag{1}$$

Once Bob decodes the file, he immediately broadcasts an acknowledgment for the whole transfer, causing the relays to stop their transmissions.

[1] The use of random coefficients ensures that the generated packets p'_i are independent with exponentially high probability [10].

3.2.2 Example 2: Multicast

Consider the example in Fig. 2.5(b), where the source wants to multicast a video stream to nodes D_1 and D_2. Say that the source transmits packets p_1 and p_2 to the router R, which broadcasts the two packets to the destinations. Since wireless receptions at different nodes are highly independent [22, 23], it is possible that when R broadcasts the two packets, node D_1 receives only p_1, while D_2 receives only p_2. In this case, node R has to retransmit both packets to allow both destinations to recover their corresponding losses. But, if node R is allowed to code, then it can XOR the two packets (i.e., $p_1 \oplus p_2$), and broadcast the XOR-ed version on the wireless medium. This single transmission allows both destinations to recover their corresponding losses, providing efficient reliability.

3.2.3 The Protocol

MORE sits below the IP layer and above the 802.11 MAC. It provides *reliable* file transfer. It is particularly suitable for delivering files of medium to large size (i.e., 8 or more packets).

(a) Source: The source breaks up the file into batches of K packets. These K uncoded packets are called *native packets*. When the 802.11 MAC is ready to send, the source creates a random linear combination of the K native packets in the current batch and broadcasts the coded packet. A *coded packet* is $p'_j = \sum_i c_{ji} p_i$, where the c_{ji}'s are random coefficients picked by the node, and the p_i's are native packets from the same batch. We call $\vec{c}_j = (c_{j1}, \ldots, c_{ji}, \ldots, c_{jK})$ the *code vector* of packet p'_j. Thus, the code vector describes how to generate the coded packet from the native packets.

The sender attaches a header to each data packet, in which it reports the packet's code vector (which will be used in decoding), the batch ID, the source and destination IP addresses, and the list of nodes that could participate in forwarding the packet. The sender includes in the forwarder list nodes that are closer to the destination than itself, ordered according to their proximity to the destination (where distances can be computed using the ETX metric [7]).

The sender keeps transmitting coded packets from the current batch until the batch is acknowledged by the destination, at which time the sender proceeds to the next batch.

(b) Forwarders: Nodes listen to all transmissions. When a node hears a packet, it checks whether it is in the packet's forwarder list. If so, the node checks whether the packet contains new information, in which case it is called an *innovative packet*. Technically speaking, a packet is innovative if it is linearly independent from the packets the node has previously received from this batch.

If the node is in the forwarder list, the arrival of this new packet triggers the node to broadcast a coded packet. To do so, the node creates a random linear combination of the coded packets it has heard from the same batch and broadcasts it. Note that *a linear combination of coded packets is also a linear combination of the corresponding native packets.* In particular, assume that the forwarder has heard coded packets of the form $p'_j = \sum_i c_{ji} p_i$, where p_i is a native packet. It linearly combines these coded packets to create more coded packets as follows: $p'' = \sum_j r_j p'_j$, where r_j's are random numbers. The resulting coded packet p'' can be expressed in terms of the native packets as follows $p'' = \sum_j (r_j \sum_i c_{ji} p_i) = \sum_i (\sum_j r_j c_{ji}) p_i$; thus, it is a linear combination of the native packets themselves.

(c) Destination: For each packet it receives, the destination checks whether the packet is innovative, i.e., it is linearly independent from previously received packets. The destination discards non-innovative packets because they do not contain new information. Once the destination receives K innovative packets, it decodes the whole batch (i.e., it obtains the native packets) using the matrix inversion in Equation (1). As soon as the destination decodes the batch, it sends an acknowledgment to the source to allow it to move to the next batch. ACKs are sent using best path routing, which is possible because MORE uses standard 802.11 and co-exists with shortest path routing.

3.2.4 Empirical Results

Figure 2.6, from [5], shows testbed results comparing MORE with two multicast protocols. The first protocol, referred to as Srcr, distributes the traffic to the multicast destinations along the edges of a multicast tree rooted at the source. The second protocol, called ExOR, exploits the broadcast nature of the medium to deliver a packet to multiple nodes simultaneously. The figure plots the average per-destination throughput in a

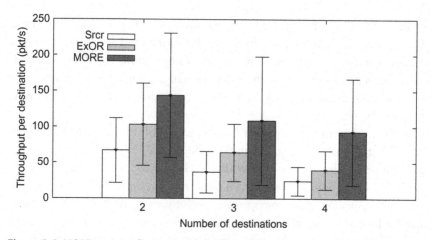

Figure 2.6 MORE, an intra-flow network coding protocol, improves multicast through-put. The figure shows the per-destination multicast throughput of MORE, ExOR, and Srcr. The thick bars show the average per-destination throughput taken over 40 runs with different nodes. The thin lines show the standard deviation.

20-node wireless testbed. In this scenario, the source multicasts a file to a varying number of destinations. As expected, the per–destination average throughput decreases with an increased number of destinations. Interestingly, however, the figure shows that MORE's throughput gain increases with an increased number of destinations. Intra-session network coding, and specifically MORE, has 35–200% throughput gain over ExOR, and 100–300% gain over Srcr. Note that these throughput gains are a result of increased reliability, which means that fewer transmissions are required to deliver a packet to its destination. The MORE paper also shows significant empirical gains for unicast traffic [5].

3.3. Analog Network Coding: Embracing Wireless Interference

Wireless networks strive to avoid scheduling multiple transmissions at the same time, in order to prevent interference. Analog network coding (ANC) adopts the opposite approach; it exploits the interference of strategically picked senders to increase network throughput. When multiple senders transmit simultaneously, the packets collide. But looking deeper at the signal level, a collision of two packets means that the channel adds their physical signals after applying attenuations and time and phase shifts. Thus, if the receiver knows the content of the packet that interfered with the packet it wants, it can cancel the signal corresponding to the known packet after correcting for the channel's effects. The receiver is left with the signal corresponding to the packet it wants, which it decodes using standard methods. This can be illustrated using the following examples.

(a) Flows Intersecting at a Router: Consider again the Alice-Bob scenario in Fig. 2.7. We explained that intersession network coding reduces the number of transmission time slots from 4 to 3. The freed slot can be used to send new data, improving wireless throughput. But, can we reduce the time slots further? Can we deliver both packets in 2 time slots?

The answer is "yes". Alice and Bob could transmit their packets simultaneously, allowing their transmissions to interfere at the router. This consumes a single time slot. Due to interference, the router receives the sum of Alice's and Bob's signals, $s_A(t) + s_B(t)$ (modulo attenuation and phase shifts). This is a collision and the router cannot decode the bits. The router, however, can simply amplify and forward the received interfered signal at the physical layer itself without decoding it. This consumes a second time slot. Since Alice knows the packet she transmitted, she also knows the signal $s_A(t)$ corresponding to her packet. She can therefore subtract $s_A(t)$ from the received interfered signal to get $s_B(t)$, from which she can decode Bob's packet. Bob can similarly recover Alice's packet. This approach is called *analog network coding*. It is analogous to digital network coding but is done over physical signals in the wireless channel itself. As a result, analog network coding reduces the required time slots from 4 to 2, doubling the wireless throughput.

(b) Flows in a Single Direction: The above is an example of inter-session analog network coding. Analog network coding, however, is also applicable to packets from the same session. Consider the chain topology in Fig. 2.8(a), where a single flow traverses 3 hops. The traditional routing approach needs 3 time slots to deliver every packet from source to destination. Digital network coding cannot reduce the number of time slots in this scenario, but analog network coding can.

Analog network coding improves the throughput of the chain topology in Fig. 2.8(a) because it allows nodes N_1 and N_3 to transmit simultaneously and have their packets received correctly despite collisions. In particular,

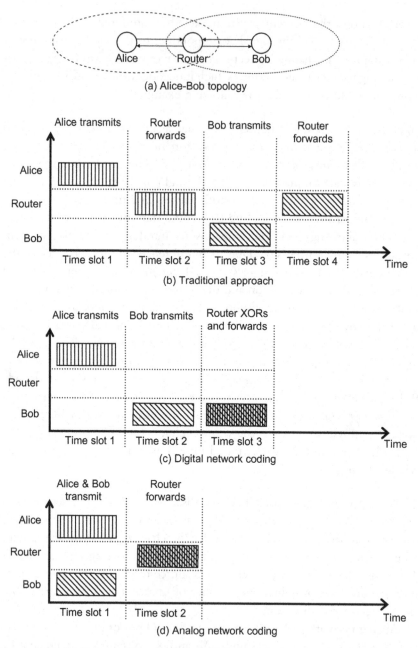

(a) Alice-Bob topology

(b) Traditional approach

(c) Digital network coding

(d) Analog network coding

Figure 2.7 Alice-Bob topology: Flows intersecting at a router. With analog network coding, Alice and Bob transmit simultaneously to the router, and the router relays the interfered signal to Alice and Bob, who decode each other's packets. This reduces the number of time slots from 4 to 2, doubling the throughput.

let node N_2 transmit packet p_i to N_3. Then, N_1 transmits the next packet p_{i+1}, whereas N_3 forwards p_i to N_4. These two transmissions happen concurrently. The destination, N_4, receives only p_i because it is outside the radio range of node N_1. But, the two packets collide at node N_2. With the traditional approach, N_2 loses the packet sent to it by N_1. In contrast, with analog network coding, N_2 exploits the fact that it knows the data in N_3's transmission because it forwarded that packet to N_3 earlier. Node N_2 can recreate the signal that N_3 sent and subtract that signal from the

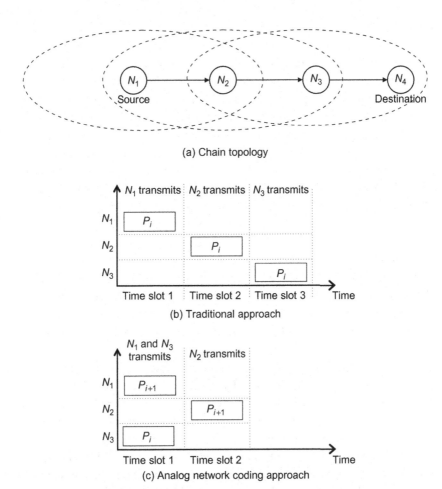

(a) Chain topology

(b) Traditional approach

(c) Analog network coding approach

Figure 2.8 Chain topology: Flows in one direction. Nodes N_1 and N_3 can transmit at the same time. N_2 gets an interfered signal, but can recover N_1's packet because it already knows the content of the interfering signal sent by N_3. This reduces the time slots to deliver a packet from 3 to 2, producing a throughput gain of $3/2 = 1.5$.

received signal. After subtraction, N_2 is left with the signal transmitted by N_1, which it can decode to obtain packet p_{i+1}. Thus, instead of requiring a time slot for transmission on each hop, we can transmit on the first and third hops simultaneously, reducing the time slots from 3 to 2. This creates a throughput gain of $3/2 = 1.5$.

Note that the chain scenario presents a form of intra-session analog network coding, whereas the Alice-Bob scenario is a case for inter-flow analog network coding.

3.3.1 The ANC Decoder

We describe the intuition underlying the ANC decoder and refer the reader to [14] and [9] for the details. Consider again the simple scenario in Fig. 2.1. Given the interfered signal, how does Alice decode Bob's packet?

First, we note that the interfered signal is not simply the sum of Alice's signal and Bob's signal, $s_A(t) + s_B(t)$. The two channels traversed by the two signals apply attenuation and phase shifts, and the two transmitters do not start at exactly the same time, i.e., the signal that Alice receives is $f_1(s_A(t)) + f_2(s_B(t-T))$, where $f_1()$ and $f_2()$ refer to the channel functions, and T is the time shift between the two signals. So to decode, Alice has to discover the channel function, $f_1()$, and apply it to her transmitted signal before subtracting. Once she does so, she is left with Bob's signal, $f_2(s_B(t-T))$, which is no different from having a single transmitter. Thus, Alice can then decode using standard methods.

So, how does Alice discover her channel function? The answer is simple; given that the two transmissions are not completely aligned, there are a few bits of Alice's signal that start before Bob's signal and do not suffer interference. Alice can use these bits to discover her channel function (using standard methods).

Bob decodes in the same way, except that since his packet starts after Alice's packet, the interference-free bits from his packets are at the end. So Bob, receives all digital samples, uses the few interference-free samples from his signal to discover his channel, compensates for his channel and subtracts his signal to obtain Alice's signal.

The above is a simplified decoder. For a more general decoder, we refer the reader to ZigZag decoding [9], which can decode analog network coding regardless of the used modulation or the packet size.

4. CONCLUSION

Network coding enables more efficient, scalable, and reliable wireless networks. These opportunities come with a need for rethinking our MAC, routing, and transport protocols. Recent years have seen significant successes in integrating network coding into wireless systems and the emergence of practical implementations with significant throughput and reliability gains. We believe that future research will continue along these lines and present novel ways for harnessing network coding in wireless systems.

REFERENCES

[1] D. Aguayo, J. Bicket, S. Biswas, G. Judd, and R. Morris. Link-level measurements from an 802.11 b mesh network. In *ACM SIGCOMM*, 2004.

[2] R. Ahlswede, N. Cai, S. R. Li, and R. W. Yeung. Network information flow. In *IEEE Transactions on Information Theory*, July 2000.

[3] S. Biswas and R. Morris. Opportunistic routing in multi-hop wireless networks. *ACM SIGCOMM*, 2005.

[4] N. Cai and R. W. Yeung. Secure Network Coding. In *ISIT*, 2002.

[5] S. Chachulski, M. Jennings, S. Katti, and D. Katabi. Trading structure for randomness in wireless opportunistic routing. In *ACM SIGCOMM*, 2007.

[6] Christos Gkantsidis and Pablo Rodriguez. Network coding for large scale content distribution. In *IEEE INFOCOM*, 2005.

[7] D. S. J. De Couto, D. Aguayo, J. Bicket, and R. Morris. A high-throughput path metric for multi-hop wireless routing. In *ACM MobiCom '03*, San Diego, California, September 2003.

[8] C. Fragouli, J. WIdmer, and J. LeBoudec. Efficient broadcasting using network coding. *IEEE/IACM Transactions on Networking*, 2007.

[9] S. Gollakota and D. Katabi. ZigZag decoding: Combating hidden terminals in wireless networks. In *SIGCOMM*, 2008.

[10] T. Ho, R. Koetter, M. Médard, M. Effros, J. Shi, and D. Karger. A random linear network coding approach to multicast. *IEEE Transactions on Information Theory*, 52(10), 4413–4430, October 2006.

[11] S. Jaggi, M. Langberg, S. Katti, T. Ho, D. Katabi, M. Médard, and M. Effros. Resilient network coding in the presence of byzantine adversaries. *IEEE Transactions on Information Theory*, 2006.

[12] S. Jaggi, P. Sanders, P. A. Chou, M. Effros, S. Egner, K. Jain, and L. Tolhuizen. Polynomial time algorithms for multicast network code construction. *IEEE Transactions on Information Theory*, 2003.

[13] A. Jiang. Network coding for joint storage and transmission with minimum cost. In *ISIT*, 2006.

[14] S. Katti, S. Gollakota, and D. Katabi. Embracing wireless interference: Analog network coding. In *ACM SIGCOMM*, 2007.

[15] S. Katti, D. Katabi, H. Balakrishnan, and M. Médard. Symbol-level network coding for wireless mesh networks. In *ACM SIGCOMM*, 2008.

[16] S. Katti, D. Katabi, W. Hu, H. S. Rahul, and M. Médard. The importance of being opportunistic: Practical network coding for wireless environments, Johns Hopkins University, Department of Computer Science, 2005.

[17] S. Katti, H. Rahul, D. Katabi, W. Hu, M. Médard, and J. Crowcroft. XORs in the air: Practical wireless network coding. In *ACM SIGCOMM*, 2006.

[18] R. Koetter and M. Médard. An algebraic approach to network coding. *IEEE/ACM Transactions on Networking*, 2003.

[19] S. R. Li, R. W. Yeung, and N. Cai. Linear network coding. In *IEEE Transactions on Information Theory*, 2003.

[20] J. Liu, D. Goeckel, and D. Towsley. Bounds on the throughput gain of network coding in unicast and multicast wireless networks. *IEEE J. Sel. A. Commun.*, 27(5), 2009.

[21] D. S. Lun, M. Médard, and R. Koetter. Efficient operation of wireless packet networks using network coding. In *International Workshop on Convergent Technologies (IWCT)*, 2005.

[22] A. K. Miu, H. Balakrishnan, and C. E. Koksal. Improving Loss Resilience with Multi-Radio Diversity in Wireless Networks. In *11th ACM MOBICOM Conference*, Cologne, Germany, September 2005.

[23] C. Reis, R. Mahajan, M. Rodrig, D. Wetherall, and J. ZahOljan. Measurement-based models of delivery and interference. In *SIGCOMM*, 2006.

[24] S. Zhang, S. Liew, and P. Lam. Physical layer network coding. In *Proc. of ACM MOBICOM 2006, Los Angeles*, USA.

[25] S. Sengupta, S. Rayanchu, and S. Banerjee. An analysis of wireless network coding for unicast sessions: The case for coding-aware routing. In *Proc. of IEEE INFOCOM*, pages 1028–1036, 2007.

[26] J. Widmer and J-Y. L. Boudec. Network coding for efficient communication in extreme networks. In *ACM SIGCOMM Workshop on DTNs*, 2005.

[27] Y. Wu, P. A. Chou, and S. Y. Kung. Information exchange in wireless networks with network coding and physical-layer broadcast. *MSR-TR-2004-78*.

[28] Y. Wu, P. A. Chou, and S.-Y. Kung. Minimum-energy multicast in mobile *ad hoc* networks using network coding. *IEEE Transactions on Communications*, 2007.

Network Coding for Content Distribution and Multimedia Streaming in Peer-to-Peer Networks

Chen Feng and **Baochun Li**
Department of Electrical and Computer Engineering, University of Toronto, Toronto, Canada

Contents

Abstract

Peer-to-peer (P2P) networks have been one of the most promising platforms to realize the potential of network coding, since end hosts (referred to as peers) at the edge of the Internet have abundant computational resources with modern processors. In this chapter, we take a journey into the application world of network coding in P2P networks, with a focus on two important applications: *content distribution* and *multimedia streaming*. For each application, we explore the possible design space of P2P systems with network coding, and provide an intuitive explanation for the advantages of using

Network Coding. DOI: 10.1016/B978-0-12-380918-6.00003-2
Copyright © 2012 Elsevier Inc. All rights reserved.

the network coding technique. We further unfold our journey through a discussion of several theoretical results and practical issues.

Keywords: Peer-to-peer networks, content distribution, multimedia streaming, network coding, playback.

1. P2P CONTENT DISTRIBUTION WITH NETWORK CODING

P2P content distribution has become increasingly popular in current-generation content distribution protocols. The basic idea in P2P content distribution protocols is surprisingly simple. Consider a single server distributing a file (usually hundreds of megabytes or even gigabytes) to a large number of end hosts (peers) over the Internet. Instead of uploading the file to every individual peer, the server first divides the file into r data blocks, and then distributes these data blocks in an efficient manner by letting participating peers exchange them with one another.

The essential advantage of P2P content distribution is to dramatically reduce the file downloading time for each peer. Intuitively, as participating peers contribute their own upload bandwidth to serve one another, the aggregate upload bandwidth in the system is significantly increased, leading to a much faster file distribution process.

1.1. How can Network Coding be Applied to P2P Content Distribution?

BitTorrent [1] has evolved into one of the most popular P2P content distribution protocols. A detailed description of the BitTorrent system can be found in [2]. In BitTorrent, when a newly arrived peer joins the system, it contacts a central *tracker* to obtain a list of some participating peers that are currently downloading the file. Typically, the tracker provides 50 peers chosen at random from among all participating peers. The new peer then attempts to establish and maintain connections to about 40 peers, which become its *neighbors*. If a peer fails to maintain at least 20 neighbors (due to peer departures, for example), it contacts the tracker again to receive a list of additional peers. In this way, a dynamic logical network—namely a P2P *overlay* network—is formed by all participating peers that are currently downloading the file.

Participating peers in BitTorrent cooperate to download the file using *swarming* techniques. The file is divided into *r* equal-sized data blocks $\{b_1, \ldots, b_r\}$. Each peer exchanges data blocks with its neighbors, until it has obtained all *r* data blocks and can depart the system. Note that the server is also a neighbor of a limited number of peers. After a peer downloads a new data block, it informs all its neighbors, so that every peer in the system knows the block availability information among its neighbors. When requesting a block from a particular neighbor, a peer typically asks for a *local rarest* block, that is, a block that is least common among all its neighbors. The purpose here is to ensure data blocks are propagated almost uniformly through the overlay network. Otherwise, with some very rare blocks in the system, the downloading time for each peer may be greatly affected by such an information bottleneck.

In general, a participating peer in any P2P content distribution system has to answer the following question when requesting data blocks: *which blocks should be downloaded, and from which neighbors?* Referred to as the *block scheduling* problem, this question needs to be addressed in an efficient and distributed fashion, with local information only. BitTorrent employs the *local rarest first* strategy, hoping to avoid information bottlenecks. With the help of network coding, however, this question can be solved in a surprisingly simple and effective manner.

Avalanche [3, 4] represents one such P2P content distribution protocol with network coding. In Avalanche, the file is again divided into *r* equal-sized data blocks. This time, each block b_i $(i = 1, \ldots, r)$ is regarded as a fixed-length vector over a finite field $\mathbb{F}_q$. Rather than uploading original data blocks to its neighbors, a peer (or the server) uploads coded blocks, where each coded block is a random linear combination of the coded blocks already received by the peer. With this change, there is no need for a peer to request specific data blocks. Instead, a peer in the system blindly downloads coded blocks from its neighbors, until it is able to reconstruct the original file.

Let us use a simple example in [5] to better illustrate how the system works. At the beginning of the file distribution process, a peer (say peer *A*) contacts the server and receives a number of coded blocks. For example, the server uploads two coded blocks x_1 and x_2 to peer *A*, where:

$$x_i = \alpha_1^i b_1 + \alpha_2^i b_2 + \cdots + \alpha_r^i b_r, \; i = 1, 2 \qquad (1)$$

with α_j^i $(1 \leq j \leq r)$ being chosen randomly from the field $\mathbb{F}_q$. When peer A needs to serve a neighboring peer (say peer B), it simply generates a coded block x_3:

$$x_3 = \alpha_1^3 x_1 + \alpha_2^3 x_2, \tag{2}$$

where α_1^3 and α_2^3 are again randomly chosen from $\mathbb{F}_q$. Substituting Equation (1) into Equation (2), we obtain:

$$x_3 = \sum_{j=1}^{r} \left(\alpha_1^3 \alpha_j^1 + \alpha_2^3 \alpha_j^2 \right) b_j.$$

It means that x_3 (and in general every coded block in the system) is some random linear combination of the original blocks $\{b_1, \ldots, b_r\}$, and the associated vector $(\alpha_1^3 \alpha_1^1 + \alpha_2^3 \alpha_1^2, \ldots, \alpha_1^3 \alpha_r^1 + \alpha_2^3 \alpha_r^2)$ is called the *global encoding vector* for x_3. Continuing this process, a peer in the system is able to collect r linearly independent coded blocks. How can a peer recover the original file from these coded blocks? As explained in Chapter 1, the global encoding vector is attached to each block as a "header", and this information is used by a peer to reconstruct the original file as long as it has received r linearly independent coded blocks.

1.2. Why is Network Coding Helpful in P2P Content Distribution?

What are the potential benefits of network coding for P2P content distribution? Recall that the difficulty of P2P content distribution is finding an optimal block scheduling algorithm, which should minimize the file downloading time in a distributed manner. This becomes even more challenging when the overlay network is changing dynamically.

Without network coding, each peer has to decide which blocks to download from which neighbors, based on local information only. This may be suboptimal due to lack of a global view, since local rarest blocks may not be global rarest blocks. In contrast, with network coding, all coded blocks are almost equally useful to any peer, and there is no need to locate and request global rarest blocks in the system. As such, an information bottleneck is avoided, which in turn reduces the file downloading time.

Another important benefit of network coding is the robustness to peer departures. Without network coding, it is possible that a few data blocks are lost, due to departures of the server as well as the peers who have these data blocks. In this unfortunate event, the original file cannot be reconstructed by the remaining peers. On the other hand, with network coding, the risk of losing certain data blocks is no longer a concern. Intuitively, since data blocks are mixed together, each of them is spreading to a large number of coded blocks in the system.

In summary, the use of network coding solves the block scheduling problem in a surprisingly simple and effective manner, leading to a shorter file downloading time. Moreover, it provides desirable robustness to peer departures. There is no need to worry about losing certain data blocks any more. All of these contribute to the usefulness of network coding in P2P content distribution. In the following, we will present a number of theoretical results to substantiate these claimed advantages of network coding.

1.3. Theoretical Results on P2P Content Distribution with Network Coding

Let us begin with the system model for P2P content distribution. The overlay network formed by the single server and participating peers can be modeled by a directed graph $G = (V, E)$, where $s \in V$ denotes the server, a vertex $v \in V - \{s\}$ corresponds to a participating peer, and every edge $e \in E$ corresponds to an overlay connection from one peer to another.[1]

This model is well suited for *static* P2P content distribution systems without peer arrivals and departures. However, peers may join and leave the system at any time in the file distribution process. To model such peer dynamics as well as block transmissions, the *trellis graph technique* [5, 6] can be applied as follows. For a directed graph $G = (V, E)$, we construct a new trellis graph $G^* = (V^*, E^*)$ with the node set:

$$V^* = \big\{ (i, t) : i \in V \text{ and } t \in \{t_0, t_1, t_2, \ldots\} \big\},$$

---

[1] If peer i maintains a TCP connection with another peer j, then we say there is an edge from peer i to peer j. Note that an edge in G is not a physical communication link; instead, it is an abstract link that consists of a path of underlying physical links.

where the node $(i, t) \in V^*$ corresponds to the node $i \in V$ at time t, and the set $\{t_0, t_1, t_2, \ldots\}$ denotes all starting times and ending times for events of peer dynamics and block transmissions. The edge set E^* is determined by the strategy adopted by the server, as well as by all the other nodes in V to request and upload coded blocks. Specifically, there are two types of edges in the trellis:

1. An edge $e \in E^*$ is added in the trellis graph from node $(i, t_k) \in V^*$ to node $(j, t_l) \in V^*$, where $(i, j) \in E$ and $t_k < t_l$, if a coded block is uploaded from node i to node j, starting at time t_k and ending at time t_l. This type of edge represents transmissions of coded blocks between neighboring peers.
2. Assume that node i joins the system at time t_k and leaves the system at time t_l. Then there is an edge with infinite capacity from node (i, t_m) to node (i, t_{m+1}), for all m satisfying $k \leq m \leq l - 1$. This type of edge represents the accumulation of received information at node i over time.

The trellis graph model presented here is a generalization of that used in [5, 6]. An illustration of the trellis graph G^* up to $t \leq t_8$ is given in Fig. 3.1, where the edges with capacity ∞ are lightened. Note that the trellis graph G^* is always acyclic, regardless of whether the directed graph G is acyclic or not. The analysis in the sequel is a direct extension of that in [5].

At time t_0, the server s is ready to distribute the file consisting of r data blocks via swarming techniques with network coding as described in Section 1.1. This can be exactly modeled as the node $(s, t_0) \in V^*$ multicasting r data blocks to all other nodes in G^* via random linear network coding.

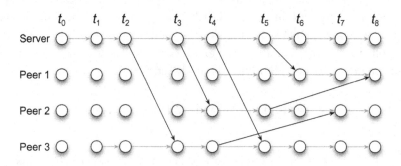

Figure 3.1 An illustration of the trellis graph G^*.

We are now in a position to determine the downloading time for a particular node $i \in V$. Consider the maximum flow from $(s, t_0) \in V^*$ to a node $(i, t_l) \in V^*$ in the trellis G^*, denoted by $f_i(t_l)$. When $f_i(t_l) \geq r$, the node (i, t_l) can recover the whole file with probability close to 1 under random linear network coding, provided that the field size q is sufficiently large (please refer to the discussion in Chapter 1 for details). Let $m(i)$ be the minimum value of l such that $f_i(t_l) \geq r$, that is, $m(i) = \min\{l : f_i(t_l) \geq r\}$. Then with high probability, the downloading time for node $i \in V$ is given by $t_{m(i)}$. Note that $t_{m(i)}$ is also a lower bound on the downloading time based on the information theoretic cut-set bound. This implies that the use of random linear network coding can achieve the minimum downloading time for this system. Even if the rare event indeed happens that node $i \in V$ cannot reconstruct the original file at time $t_{m(i)}$ due to linear dependence, it can eventually recover the file after receiving a few additional coded blocks. This can be proved by using a simple probabilistic argument.

The above analysis demonstrates the first major advantage offered by network coding—the peers do not need to decide which data blocks are to be downloaded, while still achieving the minimum downloading time for any given strategy of requesting and uploading coded blocks. We next turn to the second major advantage of network coding—robustness to peer departures.

With some peers or even the server leaving the system, it is natural to ask whether the remaining peers are able to reconstruct the original file and finish their downloads. This important question can again be answered by using the trellis graph techniques as above. For any given time t_l, assume that we are interested in whether the remaining peers in the system can recover the original file, provided that no more peers leave the system after time t_l. Toward this end, let $R(t_l) = \{(i, t_l) : i \in V\}$ and let $f_R(t_l)$ denote the maximum flow from the node $(s, t_0) \in V^*$ to the set of nodes $R(t_l)$ in the trellis graph. Then the remaining peers can recover the whole file with high probability, provided that $f_R(t_l) \geq r$ [7]. Notice that when $f_R(t_l) < r$, it is impossible for the remaining peers to recover the whole file even with the help of global information, due to the information theoretic cut-set bound. This implies that the use of network coding indeed provides the best possible robustness. The robustness issue also appears in peer-to-peer storage systems which provide reliable access to data through unreliable peers. The use of network coding offers similar benefits as shown in [8, 9].

Thus we see that the use of network coding in P2P content distribution achieves both minimum downloading time and maximum robustness with respect to the strategy of requesting and uploading coded blocks. For a given strategy, this is almost the best possible performance one can expect. However, one still needs to design a good strategy for peers to request and upload coded blocks. In fact, this question is closely related to fairness issues in real-world P2P content distribution, since a peer naturally prefers to help those neighbors that provide the best download rates. As such, a "tit-for-tat" strategy is used in Avalanche [4] for requesting and uploading coded blocks, which aims to encourage cooperation and reduce free-riding.

On the other hand, when all participating peers are operated by the same company, fairness is not a major concern, since peers are willing to cooperate with each other even without incentive. Can network coding help in this case? Deb et al. [10] have shown that with network coding, even a naive strategy of requesting and uploading coded blocks can achieve a shorter downloading time. More specifically, they consider a system model consisting of n participating peers and r data blocks to start with. Initially, every peer has only one out of the r data blocks, and each data block is equally spread in the system. The goal is to distribute all r data blocks among all the peers as fast as possible. For simplicity, they assume that time is divided into slots and slots are synchronized at the various peers. During each time slot, each peer requests coded blocks from all the neighboring peers. Then each peer chooses a neighbor uniformly at random from those who sent requests, and uploads a coded block using random linear network coding. In other words, a peer simply employs a random strategy to upload a coded block per time slot. The following theorem characterizes the performance of this strategy with network coding.

Theorem 1 ([10]). *Suppose the underlying overlay network is fully connected. Suppose the field size $q \geq \max\{r, \ln(n)\}$. Let T_b be the random variable denoting the broadcasting time (the time required by all the peers to download all data blocks) under the system model and the strategy described as above. Then:*

$$T_b \leq 5.96r + O\left(\sqrt{r \ln(r)} \ln(n)\right), \text{ with probability } 1 - O\left(\frac{1}{n}\right).$$

Further, let T_i be the downloading time for peer i, then:

$$E[T_i] \leq 5.96r + O\left(\sqrt{r\ln(r)}\ln(n)\right).$$

Observe that when $r = \Theta(n)$, it takes at least $\Theta(n)$ time slots for all the peers to download all data blocks. Thus, Theorem 1 implies that this simple strategy with network coding is order-optimal for $r = \Theta(n)$. They also prove that without network coding, the simple strategy performs strictly worse in terms of the broadcasting time. In other words, the use of network coding has the potential to improve system performance in an environment with full cooperation when $r = \Theta(n)$. Furthermore, their simulation results demonstrate that the benefits of network coding carry over to the case when r is small.

Previous theoretical results suggest that the use of network coding helps to reduce the broadcasting time, which corresponds to the maximum downloading time among all the peers. It might be natural to ask: what about other functions of downloading time, for example, the average downloading time? In fact, it is shown in [11] that, given an order at which the peers finish downloading, the use of network coding achieves any point in the optimal delay region, and in particular the average downloading time. This result is partially extended to dynamically changing network scenarios [12] in which network coding is shown to provide a robust solution that outperforms routing.

We have presented many advantages of network coding in P2P content distribution. However, these advantages do not come without a price. In fact, additional computational resources are required compared with traditional approaches. The next question naturally arises: can we reduce the computational cost without significant performance loss?

In [13], Chou *et al.* have proposed the concept of *group network coding*, in which a file is divided into equal-sized *segments* (also referred to as *generations*) with each segment further divided into equal-sized data blocks. The coding operation is performed on the blocks within the same segment, but not across different segments. Though group network coding has a potential to reduce computational complexity to a large degree, its effects on the file downloading time and robustness to peer departures are not clear at the first glance.

It has been shown in [14] that the use of group network coding is still able to achieve minimum file downloading time with high probability for any given strategy of requesting and downloading coded blocks. Moreover, the computational cost for decoding group network codes can be further reduced with a precoding technique, similar to the one used in Raptor codes (see [14] for details). In other words, a significant number of computational operations can be saved without noticeable loss in downloading time.

We next turn to the effect of group network coding on robustness to peer departures. Intuitively, the robustness degrades as the number of segments increases, since each segment contains fewer data blocks and it is likely that all of them are lost with peer departures. On the other hand, a small number of segments brings a high degree of robustness, but little saving of computational cost. In other words, there exists a fundamental robustness-complexity trade-off of network coding in P2P content distribution. It would be best if one can operate at a "sweet spot" to enjoy most robustness with reasonable computational cost. In [15], this robustness-complexity trade-off is quantitatively characterized, providing a theoretical guideline for choosing such a "sweet spot".

To summarize, network coding makes optimal use of available overlay connections, without the need for sophisticated block scheduling algorithms. At the same time, it provides best possible robustness, even if a number of peers leave the system suddenly. Moreover, if coding operation is applied within each segment rather than the whole file, the computational cost can be greatly saved without significant performance loss.

1.4. Practical Aspects of P2P Content Distribution with Network Coding

The price of network coding is mostly the computational cost, which, however, may require prohibitive computational resources. As discussed in Section 1.3, the use of group network coding can reduce the computational cost without noticeable performance loss. However, there is a trade-off between complexity and performance in general. So a question of practical interest may be: what is the maximum computational cost one may afford with modern processors? Or what are the possible configurations with respect to the number of blocks and the block size that are affordable in practice? This question has been addressed in [16] with a high-performance implementation of random linear network coding at the application layer. A number of important observations have been made in [16] with a brief summary here.

1. The number of data blocks in each segment should be less than a few hundreds with block size less than a few megabytes, in order to ensure affordable encoding and decoding.

2. The number of data blocks in each segment matters more than the block size. This suggests the use of a small number of data blocks in each segment (such as 100).

3. The optimal block size is around 2–32 KB, which provides fastest encoding and decoding. This optimal value increases with the number of data blocks in each segment.

In fact, each segment in Avalanche contains 80 data blocks, agreeing with our second observation. Each data block in Avalanche has approximately 2.3 MB, which is reasonable as suggested in our first observation. Why doesn't Avalanche further reduce the block size? This is partially due to slow connections between participating peers. With slow connections, uploading and downloading speeds become the bottleneck, rather than encoding and decoding speeds. In addition, a larger block size means less overhead in terms of the "header", which is also desirable in practice.

Another practical concern about network coding is the protection against malicious peers. In fact, even a single corrupted block—injected by a malicious peer—has the potential to pollute a large number of coded blocks and prevent participating peers from decoding. Can we detect or even correct corrupted coded blocks? In Avalanche, the use of *secure random checksums* [4] provides an on-the-fly detection of corrupted coded blocks at a low computation cost. Another solution to address malicious attacks is to use network error-correcting codes [17], which itself becomes a new research area for network coding.

2. P2P MULTIMEDIA STREAMING WITH NETWORK CODING

Peer-to-peer (P2P) multimedia streaming has recently witnessed unprecedented growth on the Internet, delivering live streaming content to millions of users in real-world applications. The essential advantage of P2P streaming is to dramatically increase the number of peers a streaming channel may sustain with dedicated streaming servers. Intuitively, as participating peers contribute their own upload bandwidth to serve one another in the same channel, the load on dedicated streaming servers is significantly mitigated.

There are a number of fundamental performance metrics that characterize "good" P2P streaming systems. Let us visit a few of them as examples. With respect to *playback quality*, if streaming content does not arrive in a timely fashion, it has to be skipped at playback, leading to a degraded playback quality. How do we consistently maintain a high playback quality at all participating peers? With respect to the *initial buffering delay*, which is experienced by a peer when it first joins or switches to a new streaming channel, how do we improve user experience with shortest possible buffering delay? With respect to *server bandwidth costs*, how do we minimize such costs by maximizing bandwidth contribution from participating peers? Last, but not the least important, how do we design a system that *scales* well to accommodate a large "flash crowd" and a high degree of peer dynamics?

These performance metrics should be given priority when evaluating a protocol that is designed specifically for P2P multimedia streaming. The playback quality and the initial buffering delay matter most to the user experience, which determines the level of user satisfaction. The server bandwidth costs, however, matter most to the companies in operation, as they directly determine most of the ongoing operational costs. In the following, we will demonstrate how the use of network coding can help to design P2P multimedia streaming systems that enjoy good overall performance with respect to all of these performance metrics.

2.1. How can Network Coding be Applied to P2P Multimedia Streaming?

With a large number of P2P multimedia streaming protocols proposed, they generally fall into two strategic categories. *Tree-based push* streaming strategies (e.g., [18]) organize participating peers into one or more multicast trees, and distribute streaming content along these trees. In contrast, *mesh-based pull* streaming strategies organize peers into a random mesh structure, with each peer having a random subset of participating peers as its neighbors. The streaming content is divided into a series of data blocks, each representing a short duration of playback. Each peer maintains a *playback buffer* that consists of data blocks to be played in the immediate future.

Every peer *periodically* exchanges block availability information of playback buffers (often referred to as *buffer maps*) with its neighbors. Based on such information, data blocks are *pulled* from appropriate neighbors, in order to meet their playback deadlines. Data blocks that are not received in time are skipped during playback, leading to degraded quality.

Compared to tree-based push strategies, mesh-based pull strategies eliminate the need to construct and maintain multicast trees, which may be difficult in practice, especially when peers join and leave the system frequently. This makes them enjoy the advantages of simplicity and robustness to peer dynamics, which are in fact inherited from the design philosophy of BitTorrent-like content distribution systems. For this reason, most real-world P2P multimedia streaming systems, such as CoolStreaming [19] and PPLive [20], are implemented using mesh-based pull strategies.

In such a streaming system, each participating peer has to decide *which data blocks to download from which neighbors*. This question is similar to the block scheduling problem in BitTorrent-like systems, but with quite a different purpose. Instead of minimizing the file downloading time, the main purpose here is to minimize the playback skips, thereby improving the playback quality. For a newly arrived peer, a short initial buffering delay is also an important concern in the design of block scheduling algorithms.

Inspired by the success of network coding in solving the block scheduling problem for content distribution systems, one may conjecture that a similar approach may work well for multimedia streaming systems as well. In fact, Wang and Li [21] have evaluated the effectiveness of applying network coding in P2P multimedia streaming, by replacing traditional block scheduling algorithms with a group network coding scheme in an experimental testbed. It has been discovered that network coding provides some marginal benefits when the overall bandwidth supply barely exceeds the demand, or when peers are volatile with respect to their arrivals and departures.

With such mildly negative results against the use of network coding, one may argue that the advantages of network coding may not be fully exploited by simply replacing block scheduling algorithms in traditional mesh-based pull protocols. This motivates a complete redesign of P2P multimedia streaming with network coding. Indeed, Wang and Li have proposed R^2 in [22]—a new streaming algorithm designed from scratch to take full advantage of network coding. Here we present a brief overview of the design principles in R^2.

2.1.1 Random Push on a Random Mesh Structure

In traditional mesh-based pull streaming protocols (henceforth referred to as PULL for brevity), the streaming content to be served is divided into a sequence of equal-sized data blocks. In R^2, the streaming content is first divided into a sequence of equal-sized segments, and each segment is further divided into k equal-sized data blocks. The coding operation is only performed within each segment, but not across different segments. This is again for the purpose of reducing the computational cost.

Suppose a peer p has received m ($m \leq k$) coded blocks in a segment s so far, denoted as $[b_1, b_2, \ldots, b_m]$. When peer p needs to serve segment s for its downstream peers, as shown in Fig. 3.2, it generates a coded block x:

$$x = \alpha_1 b_1 + \alpha_2 b_2 + \cdots + \alpha_m b_m,$$

with α_j ($1 \leq j \leq m$) being chosen randomly from the finite field $\mathbb{F}_q$. As pointed out in Section 1.1, x is ultimately some random linear combination of the original blocks in segment s. Thus a downstream peer of p is able to decode segment s as long as it has received k linearly independent coded blocks.

In PULL, a missing block on a peer is requested and pulled from an appropriate neighbor. If this block has not been received within a certain time (due to peer departures, for example), it has to be requested and pulled again, under the risk of missing a deadline. With network coding in R^2, a missing segment on a peer is served by multiple neighbors simultaneously without any explicit coordination, as illustrated in Fig. 3.3. In this way,

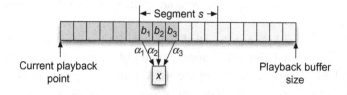

Figure 3.2 An example to illustrate the coding operation on peer p, where peer p has received 3 coded blocks within the segment s, and each segment consists of 6 blocks.

Upstream peers of peer *p*

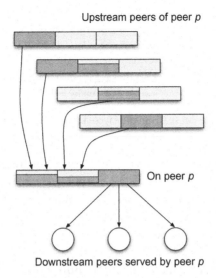

On peer *p*

Downstream peers served by peer *p*

Figure 3.3 An illustration of random push on a random mesh structure. With network coding, multiple upstream peers are able to perform push operations on coded blocks within a segment without any explicit coordination.

participating peers are able to perform *push* operations on a random mesh structure (i.e., upstream peers decide which segments to serve without any requests from downstream peers), thereby fully utilizing available overlay connections. More importantly, even if a few neighbors leave the system suddenly, other neighbors serving the same segment are still at work, with little chance of missing a deadline.

2.1.2 Timely Feedback from Downstream Peers

Before pushing coded blocks, an upstream peer should obtain precise knowledge of the missing segments on its downstream peers at any time. This requires participating peers in the system to exchange their buffer maps in a timely fashion. In PULL, buffer maps are exchanged periodically to avoid excessive overhead. While R^2 can afford "real time" exchanges of buffer maps—whenever a peer has played back or completely received a segment, it sends a new buffer map to all its neighbors.

Why is R^2 able to perform such "real time" exchanges of buffer maps? Since R^2 uses large segments instead of small data blocks (see Fig. 3.4), buffer maps that indicate segment availability information—as opposed to

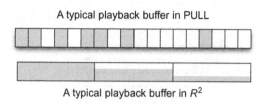

A typical playback buffer in PULL

A typical playback buffer in R^2

Figure 3.4 An illustrative comparison of playback buffers between R^2 and PULL.

block availability information—can be an order of magnitude smaller. In addition, with larger segments, it takes much longer to playback or finish downloading a segment, leading to a slower update of buffer maps. As such, with reasonable overhead, a new buffer map can be sent to all neighbors as soon as its status has been updated.

2.1.3 Synchronized Playback and Initial Buffering Delays

The use of much larger segments also makes it easier to synchronize playback buffers on different participating peers, so that all peers are playing the same segment at approximately the same time. R^2 features *synchronized playback* as follows. When a peer joins or switches to a new streaming channel, it first retrieves buffer maps from its neighbors, along with information of the current segment being played back. To synchronize the playback buffer, the new peer skips a few segments and *only* retrieves segments that are D seconds after the current playback point, where D corresponds to the *initial buffering delay*, which depends on the number of segments to be skipped and the current playback point (see Fig. 3.5). The peer then starts playback after precisely D seconds have elapsed in real time, regardless of the status of the playback buffer.

Under synchronized playback, peers are able to help each other more effectively, since their playback buffers overlap as much as possible. This desirable property is of particular importance during a flash crowd scenario, when a large number of peers join the streaming channel around the same time. As these newly arrived peers request almost the same segments, they are able to help one another as soon as they receive a small number of coded blocks. This leads to a better utilization of their upload bandwidth, which in turn improves the scalability of the streaming system.

Finally, we remark here that R^2 represents a simple design philosophy, rather than a strict protocol design. The design space of R^2 is flexible to

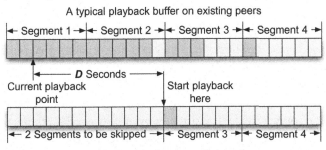

Figure 3.5 An illustration of initial buffering delays in R^2, which shows that the initial buffering delay on a newly joined peer is determined by the number of segments to be skipped and the current point of playback.

accommodate more elaborate protocols designed for different purposes. For example, a peer in R^2 has the freedom to decide which segments to be pushed to which neighbors. Referred to as a *push strategy*, this decision may be made based on timing requirements to ensure smooth playback of urgent segments, on fairness issues to encourage cooperation and reduce free-riding, or on geography considerations to reduce traffic across different ISPs.

2.2. Why is Network Coding Helpful in Multimedia Streaming?

The strict timing requirement of multimedia streaming applications has marked a significant departure from applications in content distribution. Due to such a constraint, the advantages of network coding are less obvious. In fact, it has been shown in [21] that the success story of applying network coding in content distribution cannot be simply replicated in multimedia streaming. To take full advantage of network coding, a complete redesign of P2P streaming algorithms is indeed required. This motivated the design and implementation of R^2, as described before. In this section, we intuitively explain why the use of network coding in R^2 is able to provide good *overall* performance for streaming systems.

First, with network coding, R^2 is able to use much larger segments compared to typical data blocks in PULL. The use of larger segments leads to "real time" exchanges of buffer maps without additional overhead (or even with less overhead!). With up-to-date buffer maps in R^2, participating peers are able to serve each other better. This is in contrast to PULL, where buffer maps are exchanged in a periodic fashion. As shown in [23, 24], the lack of timely exchanges of buffer maps may be a major factor that separates the actual performance of PULL from optimality.

Second, with network coding, peers in R^2 perform push operations rather than pull operations, thereby making a better use of their upload bandwidth resources. More importantly, even slow overlay connections may be utilized in R^2, which is generally impossible in PULL [23]. In short, all these factors contribute to a better utilization of peer bandwidth resources, leading to a higher playback quality and reduced server bandwidth costs.

Third, with network coding, robustness to peer departures in R^2 can be significantly enhanced. Since multiple upstream peers are serving a segment at the same time, the departure of a few of them does not constitute a challenge. In contrast, a missing block in PULL can only be served by one upstream peer at a time. Whenever an upstream peer leaves the system suddenly, the downstream peer has to find it out and request the missing block again. If this block is close to its playback deadline, the unlucky downstream peer is indeed at risk of missing a deadline.

Last but not least, with network coding, R^2 scales well to accommodate a large flash crowd. Recall that the use of network coding enables synchronized playback in R^2. With playback buffers overlapping as much as possible, newly arrived peers during a flash crowd are able to help one another immediately after they have received a few coded blocks. This allows full utilization of upload bandwidth from new peers, which in turn greatly improves the scalability of streaming systems.

2.3. Theoretical Results on P2P Multimedia Streaming with Network Coding

In this section, we present a number of analytical results on the performance of P2P streaming systems with network coding, with a focus on the fundamental limits and achievable performances of R^2. For the sake of mathematical tractability, we make a few assumptions in the system model. The key notations involved are summarized in Table 3.1 for easy reference.

First, in accordance with measurement studies of existing P2P systems (e.g., [25]), we assume peer upload capacities are the only bottlenecks in the streaming system. Second, to characterize the heterogeneity in terms

Table 3.1 Key notations in the system model

U_i	Upload capacity of a class-i peer (in blocks per second).
U_p	Average upload capacity of participating peers.
U_s	Server upload capacity (in blocks per second).
R	Streaming rate (in blocks per second).
D	Initial buffering delay (in seconds).
N	Scale of a flash crowd (the number of participating peers in the system).
k	Number of data blocks in each segment.
δ	Server strength $\left(= \frac{U_s}{NU_p}\right)$.

of peer upload capacities, we adopt the two-class model in [26], in which peers in the system are broadly classified into two classes, with each class having approximately the same upload capacity[2]. We use U_p to denote the *average* upload capacity of participating peers and U_s to denote the upload capacity of a dedicated streaming server (if multiple streaming servers exist, they can be regarded as a virtual "super-server").

We are now in a position to discuss several fundamental performance limits for P2P streaming systems with network coding. First of all, we observe that *the total bandwidth consumption should be no greater than the total bandwidth supply*. This leads to the following theorem, which has been proved in [26].

Theorem 2 ([26]). *The maximum streaming rate $R_{\max}$ is given by:*

$$R_{\max} = U_p + \frac{U_s}{N},$$

where N is the number of participating peers in the system.

Second, let us consider the buffering delay for all participating peers in the system to buffer a segment. On the one hand, since a segment consists of k data blocks, at least kN block transmissions are needed for N peers to buffer a segment. On the other hand, the aggregate upload rate is upper bounded by $U_s + NU_p$, since a peer cannot serve a segment unless

---

[2] This assumption is reasonable as there are roughly two classes of peers in P2P streaming systems—high bandwidth Ethernet peers and low bandwidth DSL peers. Although only two classes are assumed here, the analysis is easily extended to the case of multiple classes in the system.

it has received at least one coded block in this segment. Thus we have the following limit on the buffering delay:

Theorem 3. *Let D_s be the random variable denoting the buffering delay for a segment (the time required by all the peers in the system to receive the segment) under the system model described as above. Then for any given push strategy:*

$$E[D_s] \geq \frac{kN}{U_s + NU_p},$$

where N is the number of participating peers in the system, and k is the number of data blocks in each segment.

Note that at least one segment should be buffered to ensure smooth playback. By Theorem 3, it takes at least $E[D_s]$ seconds in expectation for participating peers to achieve so. In other words, Theorem 3 provides a lower bound for the shortest initial buffering delay during a flash crowd.

Given the above performance limits, we shall answer the following two questions:

- What is the sufficient condition for R^2 to achieve good overall performance?
- How far from optimality is the performance of R^2?

These questions are crucial for understanding the fundamental properties and limitations of R^2. We mainly focus on the flash crowd scenario here, since it poses unique challenges in the streaming system design, as observed in various measurement studies [19, 27]. We refer readers to [28] for analysis under other peer dynamic patterns.

During a flash crowd, a large number of peers join the system within a short period of time, just after a new live event has been released. To model a flash crowd event, we assume time is *slotted* in the sense that it takes one time slot to play back a segment. We further assume that all participating peers join the system within one time slot. We emphasize here that these assumptions are not necessary, and can easily be relaxed in the analysis. We now introduce the following definitions.

Definition 1. *The scale of a flash crowd, denoted by N, is defined as the maximum number of peers joining the system during a flash crowd event.*

Definition 2. *The server strength, denoted by δ, is defined as follows:*

$$\delta = \frac{U_s}{NU_p},$$

where U_p is the average upload capacity of participating peers, and U_s is the server upload capacity.

Before we venture into theoretical analysis of R^2, we shall specify all design choices, since R^2 is not a strict protocol design, as discussed in Section 2.1. Indeed, we adopt a simple random push strategy in the analysis. More specifically, whenever a peer has a chance to serve, it chooses a partner uniformly at random from its neighbors, and uploads a coded block in the *most urgent* segment on that partner, that is, the segment closest to the playback point not yet completely received.

The following theorem establishes a sufficient condition on smooth playback at a streaming rate R during any flash crowd with scale N.

Theorem 4 ([28]). *Suppose the underlying overlay network is fully connected. Assume that the following condition holds:*

$$U_s + NU_p \geq (1+\varepsilon)NR, \tag{3}$$

where ε is given by:

$$\varepsilon = \gamma(q) + \frac{\ln(1+\delta) - \ln\delta}{k}, \tag{4}$$

and $\gamma(q)$ is a system-wide parameter denoting the fraction of linearly dependent coded blocks induced by random linear network coding (which depends on the field size q). Then R^2 is able to achieve smooth playback at a streaming rate R under a simple random push strategy described as above, when the scale N of the flash crowd is sufficiently large.

Combining Theorem 4 with theorems 2 and 3, we are able to characterize the performance gap between R^2 and the optimal streaming scheme with regard to the sustainable streaming rate and initial buffering delay.

Corollary 1. *Suppose the underlying overlay network is fully connected. Then R^2 achieves a factor of $1 + \varepsilon$ of the maximum streaming rate $R_{\max}$ under a simple random push strategy, where ε is given in Equation (4).*

Corollary 2. *Suppose the underlying overlay network is fully connected. Then R^2 achieves a factor of $2(1 + \varepsilon)$ of the minimum initial buffering delay under a simple random push strategy, where ε is given in Equation (4).*

Corollaries 1 and 2 demonstrate that R^2 is able to support a near-optimal streaming rate with short initial buffering delays during a flash crowd, even with a simple random push strategy. Here we present a concrete numerical example to better illustrate this point. As shown in [28], the parameter $\gamma(q)$ is typically in the order of 0.1% for large q ($q \geq 64$). Hence we set $\gamma(q) = 0.1\%$ in our example. We next set the server strength δ to 0.001, and the number of data blocks in each segment k to 100 in our example. Then the sustainable streaming rate R satisfies $R \geq R_{\max}/1.07$, with initial buffering delays within a factor of 2.14 of the limit.

Notice that in above theorems, the overlay network is represented by a complete graph. However, in practical steaming systems, each participating peer only maintains a limited number of neighbors. Thus, it is of great interest to investigate the impact of a restricted neighborhood. Simulation results in [28] demonstrate that a small size of neighborhood (such as 50) is good enough to enjoy good overall performance in R^2.

2.4. Practical Aspects of P2P Multimedia Streaming with Network Coding

The design philosophy of R^2 has been applied and implemented in UUSee—a large-scale operational streaming system operated by UUSee Inc. (one of the leading peer-assisted media content providers in China). With 200 Gigabytes worth of real-world traces collected and analyzed, it has been reported in [29] that the theoretical benefit of using network coding can be achieved in practice: multiple upstream peers are allowed to collaboratively serve a downstream peer, leading to minimized buffering delays and bandwidth costs. In particular, the overall performance has been satisfactory for normal-quality videos in UUSee. For high-quality videos, the buffering delay could be larger due to the high bandwidth demand. Nevertheless, the UUSee measurements suggest that the delay is in general below a reasonable 20 seconds.

Any advantages may come with trade-offs. We shall now look at the flip-side of the coin—some practical issues in R^2—to get a complete picture.

Similar to content distribution applications, the computational cost of network coding is again a major concern for multimedia streaming applications. Even with modern processors, it may not be computationally feasible to support more than a few hundred data blocks in each segment. On the other hand, the use of large segments and small blocks is a key to the success of R^2, as explained before. Therefore, one shall maximize the number of data blocks in each segment and minimize the block size, subject to practical constraints.

For example, each segment in R^2 is divided into 128 data blocks of 2 KB each. With this design choice, good overall performance has been observed, but at the cost of sustained high CPU usage. In other words, we cannot afford a larger number of data blocks in each segment due to CPU constraints. Shall we choose a smaller block size in R^2? Note that the overhead in terms of the "header" is around 6% for this design choice, when the field size $q = 256$. Thus a smaller block size may lead to excessive overhead. To summarize, there is a trade-off between performance, computational cost and overhead. One shall find the right compromise for any particular streaming system.

Another practical concern in R^2 is the *braking redundancy*. Recall that a downstream peer sends a new buffer map to all its neighbors immediately after it has completely received a segment. This buffer map is also used as a signal to stop upstream peers from serving the segment. As it takes time for the braking signal to reach upstream peers, a downstream peer may receive additional redundant blocks after a segment is complete. How shall we minimize this braking redundancy? One engineering approach is to allow an "early braking" mechanism, which encourages a downstream peer to stop a subset of its upstream peers even *before* a segment is completely received. However, the design of such an "early braking" algorithm still remains a challenge.

Finally, a possible drawback of R^2 is that the time between the occurrence of a live event and its playback is the same across the board in all participating peers due to synchronized playback. Though harmful to some applications, this may be an advantage to those involving live interaction (such as live voting with SMS): all participating peers will view the same content at the same time, such that interactive behavior starts to occur at the same time as well.

3. CONCLUSION

The main objective of this chapter is to explore the potential benefits that network coding may offer in P2P networks. In particular, we have addressed two major applications: content distribution and multimedia streaming. To achieve this goal, we first describe how network coding can be successfully used in each application. We then provide a number of key insights on the advantages offered by network coding.

- In P2P content distribution, we have shown that the use of network coding solves the block scheduling problem in a surprisingly simple and effective way, leading to a shorter file downloading time and better robustness to peer departures.
- In P2P multimedia streaming, we have shown that a complete redesign of streaming protocols is required in order to take full advantages of network coding. In particular, we have presented R^2—a new streaming system design with network coding—and explained why the use of network coding in R^2 is able to fully utilize available bandwidth resources, thereby improving the overall system performance.

To deepen understanding of these claimed advantages, we further provide a number of selected theoretical results. Finally, we present several practical issues that deserve special attention in real-world system design. We believe such an exploration could shed light on future applications of network coding in P2P networks.

REFERENCES

[1] B. Cohen. Incentives Build Robustness in BitTorrent. *Proc. of 1st Workshop on Economics of Peer-to-Peer Systems*, (Berkeley, CA), June 5–6, 2003.

[2] J. Pouwelse, P. Garbacki, D. Epema, and H. Sips. The Bittorrent P2P File-sharing System: Measurements and Analysis. *Proc. of 4th International Workshop on Peer-to-Peer Systems (IPTPS)*, (Ithaca, New York), Feb. 24–25, 2005.

[3] C. Gkantsidis and P. Rodriguez. Network Coding for Large Scale Content Distribution. *Proc. of IEEE INFOCOM 2005*, (Miami, FL), March 13–17, 2005.

[4] C. Gkantsidis, J. Miller, and P. Rodriguez. Anatomy of a P2P Content Distribution System with Network Coding. *Proc. of 5th International Workshop on Peer-to-Peer Systems (IPTPS)*, (Santa Barbara, CA), Feb. 27–28, 2006.

[5] R. W. Yeung. Avalanche: A network coding analysis. *Communications in Information and Systems*, vol. 7, pp. 353–358, 2007.

[6] Yunnan Wu. A Trellis Connectivity Analysis of Random Linear Network Coding with Buffering. *Proc. of International Symposium on Information Theory (ISIT)*, (Nice, France), June 24–29, 2007.

[7] T. Ho, R. Koetter, M. Médard, M. Effros, J. Shi, and D. Karger. A random linear network coding approach to multicast. *IEEE Trans. Inf. Theory*, vol. 52, no. 10, pp. 4413–4430, October 2006.

[8] A. G. Dimakis, P. B. Godfrey, Y. Wu, M. Wainwright, and K. Ramchandran. Network coding for distributed storage systems. *IEEE Trans. Inf. Theory*, vol. 56, no. 9, pp. 4539–4551, Sep. 2010.

[9] S. Acedanski, S. Deb, M. Médard, and R. Koetter. How Good is Random Linear Coding Based Distributed Networked Storage? *First Workshop on Network Coding, Theory, and Applications (NetCod)*, (Riva del Garda, Italy), Apr. 2005.

[10] S. Deb, M. Médard, and C. Choute. Algebraic gossip: A network coding approach to optimal multiple rumor mongering. *IEEE Trans. Inf. Theory*, vol. 52, no. 6, pp. 2486–2507, June 2006.

[11] Yunnan Wu, Y. C. Hu, J. Li, and P. A. Chou. The Delay Region for P2P File Transfer. *Proc. of International Symposium on Information Theory (ISIT)*, (Coex, Seoul, Korea), June 28–July 3, 2009.

[12] C. S. Chang, T. Ho, M. Effros, M. Médard, and B. Leong. Issues in Peer-to-Peer Networking: a Coding Optimization Approach. *Proc. of the 2010 IEEE International Symposium on Network Coding (NetCod)*, (Toronto, Canada), June 9–11, 2010.

[13] P. Chou, Y. Wu, and K. Jain. Practical Network Coding. *Proc. of Allerton Conference on Communication, Control, and Computing*, (Monticello, Illinois), Oct. 1–3, 2003.

[14] P. Maymounkov, N. J. A. Harvey, and D. S. Lun. Methods for Efficient Network Coding. *Proc. of Allerton Conference on Communication, Control, and Computing*, (Monticello, Illinois), Oct. 1–3, 2006.

[15] D. Niu and B. Li. On the Resilience-Complexity Tradeoff of Network Coding in Dynamic P2P Networks. *Proc. of 15th IEEE International Workshop on Quality of Service (IWQoS)*, (Evanston, IL), June 21–22, 2007.

[16] M. Wang and B. Li. How Practical is Network Coding?. *Proc. of 14th IEEE International Workshop on Quality of Service (IWQoS)*, (New Haven, CT), June 19–21, 2006.

[17] R. Koetter and F. R. Kschischang. Coding for errors and erasures in random network coding. *IEEE Trans. Inf. Theory*, vol. 54, no. 8, pp. 3579–3591, Aug. 2008.

[18] V. Venkataraman, K. Yoshida, and P. Francis. Chunkyspread: Heterogeneous Unstructured Tree-based Peer-to-Peer Multicast. *Proc. of 14th IEEE International Conference on Network Protocols (ICNP)*, (Santa Barbara, CA), Nov. 12–15, 2006.

[19] S. Xie, B. Li, G.-Y. Keung, and X. Zhang. Coolstreaming: Design, theory, and practice. *IEEE Transactions on Multimedia*, vol. 9, pp. 1661–1671, December 2007.

[20] Y. Huang, Z. J. Fu, D. M. Chiu, C.S. Lui, and C. Huang. Challenges, Design and Analysis of a Large-scale P2P VoD System. *Proc. of ACM Sigcomm*, (Seattle, WA), Aug. 17–22, 2008.

[21] M. Wang and B. Li. Lava: A Reality Check of Network Coding in Peer-to-Peer Live Streaming. *Proc. of IEEE INFOCOM*, (Anchorage, Alaska), May 6–12, 2007.

[22] M. Wang and B. Li. R^2: Random push with random network coding in live peer-to-peer streaming. *IEEE J. Sel. Areas Comm.*, vol. 25, no. 9, pp. 1655–1666, Dec. 2007.

[23] M. Zhang, Q. Zhang, L. Sun, and S. Yang. Understanding the power of pull-based streaming protocol: Can we do better? *IEEE J. on Sel. Areas in Communications*, vol. 25, pp. 1678–1694, Dec. 2007.

[24] C. Feng, B. Li, and B. Li. Understanding the Performance Gap between Pull-based Mesh Streaming Protocols and Fundamental Limits. *Proc. of IEEE INFOCOM*, (Rio de Janeiro, Brazil), April 19–25, 2009.

[25] L. Guo, S. Chen, Z. Xiao, E. Tan, X. Ding, and X. Zhang. Measurements, Analysis, and Modeling of BitTorrent-like Systems. *Proc. of Internet Measurement Conference (IMC)*, (Berkeley, CA), Oct. 19–21, 2005.

[26] R. Kumar, Y. Liu, and K. W. Ross. Stochastic Fluid Theory for P2P Streaming Systems. *Proc. of IEEE INFOCOM*, (Anchorage, Alaska), May 6–12, 2007.

[27] X. Hei, C. Liang, J. Liang, Y. Liu, and K. W. Ross. A measurement study of a large-scale P2P IPTV system. *IEEE Trans. Multimedia*, vol. 9, no. 8, pp. 1672–1687, Dec. 2007.

[28] C. Feng and B. Li. On Large-Scale Peer-to-Peer Streaming Systems with Network Coding. *Proc. of ACM Multimedia*, (Vancouver, BC), Oct. 27–Nov. 1, 2008.

[29] Zimu Liu, Chuan Wu, Baochun Li, and Shuqiao Zhao. UUSee: Large-Scale Operational On-Demand Streaming with Random Network Coding. *Proc. of IEEE INFOCOM*, (San Diego, California), March 15–19, 2010.

CHAPTER 4

Network Coding in the Real World

Janus Heide, Morten V. Pedersen, Frank H.P. Fitzek,
and **Torben Larsen**
Department of Electronic Systems, Aalborg University, Aalborg, Denmark

Contents

Abstract

This chapter discusses the implementation of network coding on commercial mobile platforms with a focus on mobile phones. Implementation of network coding on mobile platforms poses several major challenges owing to their limited memory, energy, and computational power. These challenges open new research directions, and lead to several theoretical and practical research problems.

Keywords: Cellular networks, *ad hoc* networks, content-distribution networks, performance evaluation, implementation issue, optimization, code design.

1. INTRODUCTION: IT'S NOT ROCKET SCIENCE

Whenever people want to underline that something is easy, they tend to use the phrase "It's not rocket science". But what is the big deal about rocket science? As long as you are familiar with the basic laws of Physics,

Network Coding. DOI: 10.1016/B978-0-12-380918-6.00004-4
Copyright © 2012 Elsevier Inc. All rights reserved.

such as conservation of momentum and Newton's laws, it is easy to derive the basic formulae to send a rocket into space. The problem is not in the theoretical work but instead lies in the real world. Even if you know the mass of your rocket and the amount of fuel, the question is how to correctly use the fuel in the rocket. Therefore the transition of knowledge from the theoretical domain to the real world is an important field, especially if the implementation can generate valuable feedback for the theoretical work.

In this chapter we focus on applications of Network Coding (NC) in mobile networks. Cellular phones constitute an exciting platform owing to their wide adoption, high availability, and high mobility. However, mobile devices have limited computational capabilities, compared to server or desktop machines. In addition, mobile phones have a limited energy supply, hence the amount of consumed energy is a major consideration in the design process. In particular, any throughput gains achievable by using NC should have a minimum penalty in terms of additional energy consumption. The approaches that have been proposed for more powerful platforms might not be applicable for mobile phones, hence new techniques need to be developed to address the specific characteristics of mobile devices.

2. NETWORK CODING FOR MOBILE PHONES

In order to motivate the applications of NC we start by presenting several scenarios that show potential applications and benefits of the NC technique for mobile phones.

Mobile phones are able to create, store, and exchange multimedia content. Currently, many mobile phones can upload the content to cloud services such as Facebook, Twitter, MobileMe, etc. Typically, the creator of the content first uploads the content and then sends a link to the content to other users. However, this might result in long delays and contribute to congestion in the uplink and downlink channels, as well as in the core network.

When a group of people are located in close proximity they might want to consume the same content, such as in the color mobile application presented in [1]. For such applications, it is desirable to distribute the content directly among mobile phones without any overlay network. Direct content delivery helps to reduce transmission delays as well as the amount of traffic at the base stations and core networks. Direct content delivery could be beneficial for distributing photos, music, videos, files, and other types of content. References [2, 3] present a system in which one mobile device sends its data directly to the neighboring devices over one hop or multi-hop mobile networks. Figure 4.1 shows video broadcasting from one device to multiple receivers. References [2, 3] advocate the use of network coding for direct content delivery in wireless devices, and show that the NC technique is instrumental for efficient implementation of such systems.

In a single hop network, the content exchange can be implemented by using unicast or broadcast connections. With unicast connections, the bandwidth and energy consumption depends on the number of receiving devices. Data broadcast is a more efficient solution in terms of energy and bandwidth, especially for a large number of users, since multiple receivers can obtain data from a single transmission. In the case of error prone communication links, some form of coding needs to be applied to recover from errors and erasures.

Figure 4.1 Video broadcasting from one device to multiple receivers.

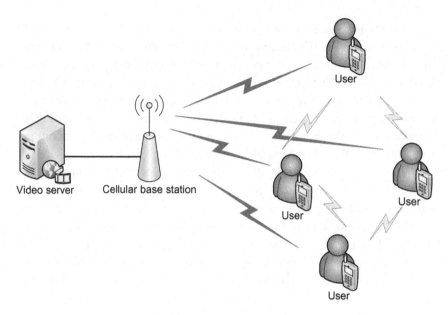

Figure 4.2 Content distribution in a Multi-Hop Network.

In addition, if the receiver nodes cooperate by forwarding packets to each other, the workload of the server can be significantly reduced. In the case where the source of the content is located far away and the communication links are likely to be error prone, cooperative downloading is beneficial (see Fig. 4.2). In this case the NC technique can be leveraged to facilitate the distribution of data among the cooperative devices. Indeed, with the standard coding techniques, the server needs to send additional data to enable nodes to recover from errors. With the network coding technique, each receiving device can forward recoded[1] information to other nodes, alleviating the load at the server.

The NC technique can also facilitate content delivery in developing countries where the cellular coverage might be sparse or unreliable.

---

[1] When a device recodes information it combines the partial information it holds, in a similar way as a source combines all information when encoding. Detailed information about recoding will be presented in Section 3.

Suppose that a user, referred to as a *spreader*, gets access to a hot-spot in a city and spreads the content on his/her device while traveling by transmitting the information to other users, referred to as *leechers*, that have no direct access to the network. By using the NC technique leechers can efficiently obtain content from different spreaders. Using NC allows the leechers to get the content from different sources without being exposed to the "coupon collector's problem". The "coupon collector's problem" is the problem of collecting all unique coupons from a set of coupons, when coupons are drawn one at a time from the complete set of coupons (with replacement). Initially, there is a high probability of drawing a unique (previously not seen) coupon, but as more and more coupons have been seen this probability decreases.

Another example is given in Fig. 4.3 where two spreaders, A and B, are conveying information to different leechers via multi-hop routes. Leechers 1, 4 and 5 are receiving packets from one single spreader, and leechers 2 and 3 are getting their packets from two spreaders. In this example NC helps to

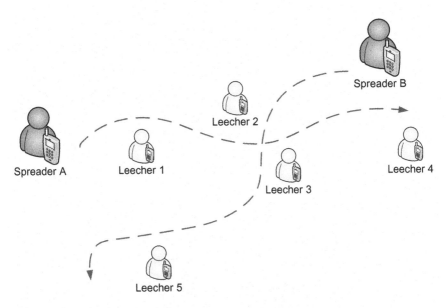

Figure 4.3 Viral content spreading.

decrease the amount of signaling between the leecher and spreader. In this particular setup leecher 2 and leecher 3 forward packets from spreader B and leecher 1 towards leecher 4 and leecher 5, respectively. Traditionally leecher 2 and 3 would need coordination in order to send information that is guaranteed to be useful for the receiving devices. With NC techniques this coordination can be almost eliminated, because the nodes simply combine whatever information they have received and send the combination as a new packet. Since this combination includes several packets, there is high probability that the receiver will be able to extract new information from the received packets.

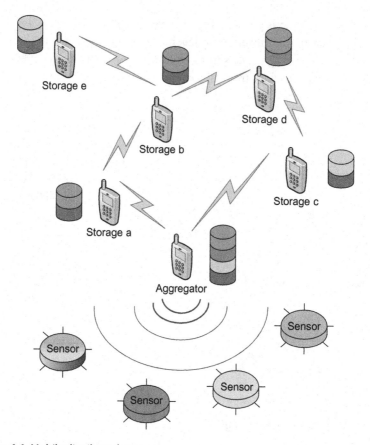

Figure 4.4 Mobile distributed storage.

Another application domain for the NC technique is distributed storage. In our previous work [4] we showed that the reliability of a sensor network could be improved by distributing the content across multiple mobile devices. The advantage of such an approach is that even if some of the devices malfunction or deplete their batteries, there are still enough devices that can send the information. Figure 4.4 depicts an example of a mobile storage network. The system has an *aggregator* that receives data from different on-board sensors and stores partial information at different mobile devices. In order to reduce the amount of storage required on the remote devices, only partial information is stored at each device. Without NC a complicated signaling protocol is required in order to ensure that the system is robust against disk failures. For example, suppose that we would like to store 100 unique information packets. With the NC technique we would create e.g., 300 coded packets and distribute them among all nodes. In case some nodes are lost, we could still decode all stored information, with high probability, as long as we can retrieve 100 different coded information packets.

As shown in [5, 6] NC is applicable for multi-path reception. The main idea is that a mobile device can add channel capacity dynamically by adding additional air interfaces. An example could be the reception of a video over a WiFi interface. In case the WiFi link is error prone or the video itself requires additional bandwidth, the mobile device can use a cellular link for additional data, as shown in Fig. 4.5. With traditional approaches, a coordination mechanism is required to guarantee performance, and to ensure that the wireless and cellular link send different packets. The NC technique can solve this problem by mixing packets at the content server.

3. SYSTEM COMPONENTS AND DESIGN CHOICES

Any NC system requires, at minimum, a protocol that governs the behavior of the system, and an implementation of the necessary coding operations. A node that is a part of this distribution system can have one of the following three roles: (1) it can be a source and thus encode and transmit data; (2) it can be a sink and thus attempt to collect the distributed data in order to decode the data; (3) it can also be a relay which holds partial data and distributes it to other relays or sinks.

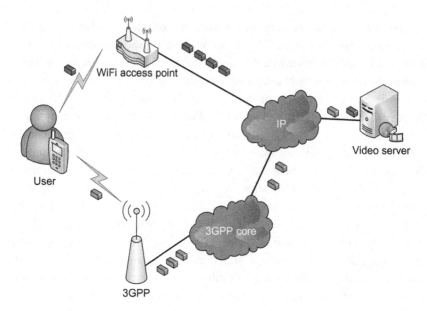

Figure 4.5 Multi-path reception.

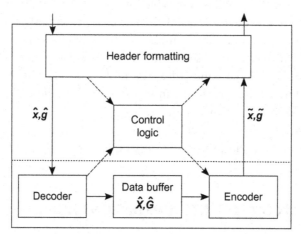

Figure 4.6 NC system components.

Figure 4.6 provides an overview of the components needed for an NC system, and their interaction. The protocol part is located above the dashed line, and contains the control block that defines how each node in the system should behave. In particular, it determines when a node should encode and send data, transmit information about its state to other nodes, or request data from another node. Actions are performed based on inputs that are either available locally or received from other nodes in the network. In order to communicate with other nodes in the system it is practical to define a header format that describes the format of packets transmitted between the nodes.[2] The coding implementation is located below the dashed line and constitutes the decoder, data buffer, and encoder blocks. In this chapter we assume that coding operations are performed over a finite field of size $q = 2^n$.

The data, of size B, that is to be transferred from a source to one or more sinks. The data can be a part of a file or a media stream. The file is divided into $\lceil \frac{B}{m} \rceil$ pieces, referred to as *symbols*. The symbols are grouped into *generations* (also referred to as *source blocks* or *batches*), each generation contains g symbols, where g is referred to as the *generation size*. In practice each symbol is represented by one or more field elements which supply the necessary field operations. Thus each symbol should be thought of as a vector of finite field elements. Note that the total size of a generation is equal to $g \cdot m$, and the total number of generations is equal to $\lceil \frac{B}{g \cdot m} \rceil$.

The g original symbols of length m in one generation, are arranged in the matrix $\boldsymbol{M} = [\boldsymbol{m}_1; \boldsymbol{m}_2; \ldots; \boldsymbol{m}_g]$, where $\boldsymbol{m}_i$ is the ith symbol in the generation. Generation number 0 constitutes the first g symbol, or the first $g \cdot m$ bytes of data.

To encode a new symbol $\boldsymbol{x}$ from a generation at the source, $\boldsymbol{M}$ is multiplied with a *coding vector* $\boldsymbol{g}$ of length g, $\boldsymbol{x} = \boldsymbol{M} \cdot \boldsymbol{g}$. In this way we can construct $g + r$ coded symbols (each symbol is associated with the corresponding coding vector), where r is the number of redundant symbols. When a coded symbol is transmitted on the network it is accompanied by its coding vector, and together they form a coded packet. For practical reasons, each coded symbol is a (linear) combination of the original symbols from one generation. The benefit is that an unlimited number of coded symbols can be created. Since the rate of the code is not fixed, it is referred to as *rateless*. In Random Linear Network Coding (RLNC) the coding vector is drawn randomly, while in deterministic approaches it is generated

[2] The detailed format is outside of the scope of this chapter.

based on some predefined algorithm. Thus to implement encoding we may need a random number generator along with an adder and a multiplier.[3]

In order for a sink to successfully decode a generation, it must receive g linearly independent symbols and coding vectors from that generation. All received symbols are placed in the matrix $\hat{X} = [\hat{x}_1; \hat{x}_2; \ldots; \hat{x}_g]$ and all coding vectors are placed in the matrix $\hat{G} = [\hat{g}_1; \hat{g}_2; \ldots; \hat{g}_g]$, referred to as the decoding matrix. The original data M can then be decoded as $\hat{M} = \hat{X} \cdot \hat{G}^{-1}$. Usually, if g or more coded symbols from a generation are received, the original data in that generation can be decoded. This is a much looser condition, compared to when no coding is used, where exactly <u>all</u> g unique original symbols must be collected.

Any node that has received $g' \geq 2$ linearly independent symbols for a generation, can create a new packet (this operation is referred to as *recoding*). All received symbols are placed in the matrix $\hat{X} = [\hat{x}_1; \hat{x}_2; \ldots; \hat{x}_{g'}]$ and all coding vectors are placed in the matrix $\hat{G} = [\hat{g}_1; \hat{g}_2; \ldots; \hat{g}_{g'}]$. To generate a new symbol $\tilde{x}$, matrix $\hat{X}$ is multiplied with a randomly generated vector h of length g' (i.e., $\tilde{x} = \hat{X} \cdot h$), and its corresponding coding vector is equal to $\tilde{g} = \hat{G} \cdot h$. This way we can construct r' randomly generated recoded symbols (each symbol is corresponding to a coding vector). It is possible to create $r' > g'$ symbols, however only g' of them will be independent. Note that h is only used locally, and that there is no need to distinguish between coded and recoded symbols. In practice this means that a node that has received more than one symbol can recombine those symbols into recoded symbols, similar to the way coded symbols are constructed at the source.

Figure 4.7 depicts an NC system and its operation over an erasure-prone channel. The encoder uses g original source symbols to create k encoded symbols. The encoded symbols are then transmitted via an unreliable channel which erases a certain number of symbols. The receiver collects enough symbols until decoding is possible (in general a little more than g symbols), and decodes the original data. A significant difference between the NC scheme and fountain codes is that it is possible to code at intermediate nodes (between the sender and receiver). In recoding an intermediate node recombines the incomplete data it has received to create new recoded symbols, which is illustrated in the bottom right corner of the figure.

---

[3] All operations are performed over a finite field.

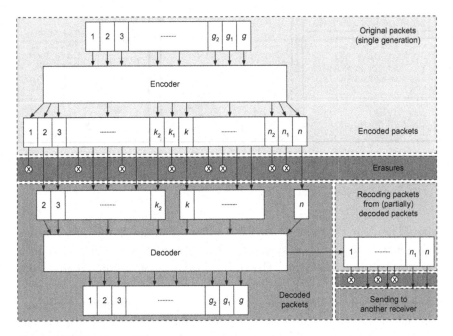

Figure 4.7 NC system with erasures.

4. PRACTICAL PROBLEMS

Most operating systems can provide "random" number generators. Unfortunately these numbers are not always sufficiently random (see Fig. 4.8). This can lead to a higher probability of linear dependency which will affect the performance of the system. The easiest way to avoid such problems is to use a random number generator that has been shown to have acceptable performance [7]. See for example [8] for an implementation of the Mersenne Twister.

In the binary extension field, addition and subtraction operations are identical to the bitwise XOR operation, which can be performed very efficiently. Multiplication and division are more complicated. One way to implement operations over higher order fields is by using look-up tables. In this case, multiplication and division operations are performed by a number of table look-ups together with some other basic operations. In [9] several different look-up approaches are presented and small code examples are provided. For these approaches to achieve good performance the tables

Figure 4.8 Dilbert on randomness. © United Feature Syndicate INC./Dist. by PIB Copenhagen 2010.

should reside in the cache of the CPU, otherwise frequent memory accesses can reduce the coding throughput. To this end, it is desirable to reduce the size of the tables, which depends on the size of the underlying finite field.

Alternatively, multiplication and division can be implemented by multiplying two polynomials (that represent symbols of a finite field), modulo an irreducible polynomial. Division can be performed with the Euclidean algorithm (see [10, p. 14,122] for details and code examples). In these algorithms, several simple operations are performed in order to calculate the result. The performance of both approaches depends on the platform on which they are deployed, in particular the relative speed of the CPU and memory access. Several additional optimization techniques can be deployed as well.

Matrix inversion can be performed with several different algorithms, the commonly used one is the Gauss–Jordan algorithm. This algorithm is simple and can be used to decode data partially, which is an advantage because packets arrive one at a time and hence the final decoding delay can be reduced. Other algorithms exist, see for example [11]. However, these often require that the matrix that is to be inverted has some specific properties, or only performs better under some conditions.

To compare different choices of methods and optimization techniques we look at the existing literature on implementing encoding and decoding operations for NC. As decoding is more computationally demanding than encoding we will focus on reported decoding throughput. We present the results in approximately chronological order to provide an overview of the recent results in the field.

5. A BINARY DETERMINISTIC APPROACH

The first mobile implementation was based on deterministic/opportunistic binary NC [12]. In such a system, NC is applied to improve the properties of a traditional broadcast network. Encoding and recoding is performed when the system detects *coding potential* by XOR-ing a relatively low number of packets together. We say that there is a coding potential when it is detected that coding can improve the network throughput. The coded packets can subsequently be decoded at other sinks. Because only few packets are coded using the XOR operation, encoding and decoding operations are not computationally demanding. This is beneficial, especially on mobile phones. In practice it can be difficult to detect coding opportunities since a network is a distributed system, and all nodes hold incomplete system state information. Thus, in such a system the coding implementation is relatively simple whereas the logic that controls it is complicated.

The COPE architecture, introduced in [13], uses deterministic/opportunistic binary inter-flow NC. The NC technique is applied in between the data link and the network layer. The system was designed for a wireless mesh network with laptop computers. This architecture cannot be applied directly for mobile phones because their network stacks are generally closed and any changes in the stack require the cooperation of the phone manufacturer. Additionally, the COPE architecture relies heavily on packet overhearing. Thus, the number of packets that must be received by each sink is relatively high, which leads to increased energy consumption. This illustrates that the systems designed for PCs and mobile phones have different performance criteria.

Reference [12] presents a system inspired by COPE, and adapted to the mobile phone domain and a different distribution scenario. The system is implemented and tested on Nokia N95 mobile phones. This NC system uses deterministic/opportunistic binary inter-flow NC on the application layer to improve content distribution from a single source to multiple cooperating sinks. A single source transmits the same content to a group of sinks via a global wireless link. To enable cooperation between the sinks, the source splits the content into different streams. This technique is typically referred to as *content-splitting*. The sinks can then cooperate by exchanging data over a local wireless link. To improve the efficiency of the local exchange, recoding is used on the sinks and applied when coding opportunities are detected.

6. RANDOM LINEAR NETWORK CODING (RLNC)

RLNC is based on dense random coding vectors over large finite fields. With RLNC, each coded packet is a combination of several packets over the selected finite field, typically of size $2^8, 2^{16}$, and 2^{32}. Fast implementation of operations over such fields is a non-trivial task. Accordingly, achieving high encoding and decoding throughput is the main challenge, especially on platforms with low computational capabilities. The first implementations targeted PCs are reported in [14–16].

In [17] a baseline implementation is presented and benchmarked using the Nokia N95 mobile phone with 332 MHz ARM 11 CPU. The implementation is based on logarithmic and exponential look-up tables. Matrix inversion is performed with a progressive version of the Gauss-Jordan algorithm. Field sizes of 2^8 and 2^{16} are used, and the tested generation size is between 10 and 400. Data is coded in blocks of 1.5 kB so that it can be transmitted via WLAN without fragmentation.

Reported results are illustrated in Fig. 4.9. The reported coding throughput is approximately 50% higher for a generation size of ten, when

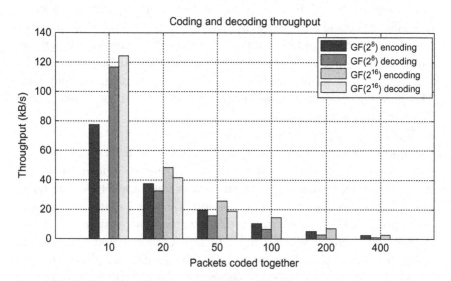

Figure 4.9 Coding throughput on a Nokia N95 mobile phone, with look-up table based implementation. Generation sizes between 10 and 400 and the block size of 1.5 kB [17] are tested.

the field size is 2^{16} compared to 2^8. For higher generation sizes the throughput increase gained by using field size of 2^{16} is very modest. The reported decoding throughput is below 20 kB/s for generation sizes above 64. Thus, the results from this baseline implementation are not very encouraging.

In [18] the system is implemented and deployed on the iPhone 3G and second generation iPod Touch, which have 412 MHz and 533 MHz ARM 11 CPU, respectively. Both an implementation based on logarithmic and exponential look-up tables and a *loop based* approach is presented. In the loop based approach multiplication is implemented by a series of basic operations. This can be an advantage in cases where the look-up table is to reside in the cache of the CPU[4]. If this is the case we may expect to get a high number of cache misses, which can significantly degrade the performance of table based approaches. Matrix inversion is performed by using the Gauss-Jordan algorithm. A field size of 2^8 is used and generation sizes between 64 and 256 are tested. Data is coded in blocks of between 128 B and 16 kB. In addition the implementation is optimized by using Single Instruction, Multiple Data (SIMD) instructions, which are shown to yield significant throughput improvements.

In this system, a decoding throughput of up to 370 kB/s is reported at a generation size of 64. The loop based approach achieves approximately twice the throughput of the look-up table approach. To obtain these results the authors have employed various tweaks, e.g., hand optimization of the compiled assembly. The authors also performed a small experiment on the power consumption by decoding and playing a stream of 77.5 kB/s. In the observed scenario NC adds approximately 50% to the power consumption.

Depending on the application, the observed values of the coding throughput may or may not be acceptable. It is worth noticing that the throughput values are reported at 100% CPU, and that most applications will use CPU cycles for jobs other than transmitting data. Secondly, a high CPU utilization will result in high energy consumption. Thus, further optimization is required for the NC technique to be applicable in a broader range of applications.

---

[4] As the field size increases, so does the size of the look-up tables. Additionally CPU's used in mobile phones generally have a much smaller cache compared to that of PC CPU's.

7. SPEEDING UP RLNC THROUGH OPTIMIZATIONS

To get a higher coding throughput, different optimization techniques can be applied. For example the use of SIMD instruction set on modern CPU, parallel execution on multi-core CPU, or dedicated hardware support for NC. SIMD instruction sets are available on most modern CPUs including those used in mobile phones. Parallel execution on multiple cores can improve throughput with a factor up to the number of available cores, compared to serial execution. Currently no mobile phone incorporates a multi-core CPU and thus parallelization does not help. However, this can change in the near future if the multi-core trend from PCs reaches the mobile phone sector. Dedicated hardware is particularly interesting as it can potentially allow for very fast coding at a reduced power consumption. However, the cost of incorporating additional hardware can be prohibitive. One solution that has been proposed in the PC domain is to use graphics processing units (GPUs). This way, an existing piece of hardware can be reused for a new task, which is an interesting and inexpensive way to implement NC.

So far only SIMD instructions have been applied to the mobile phone domain [18], whereas experiments with multi-threaded implementation and the use of GPUs are exclusive to PCs. However, these techniques may be applicable to mobile phones in the near future and thus we study the PC implementations here. Currently only a few mobile phones have a multi-core CPU, but several phones incorporate a GPU.

In [19] a parallel implementation is deployed on several platforms with the following hardware: Quad CPU Intel Pentium 4 Xeon 2.8 GHz; Dual CPU Intel Pentium 4 Xeon 3.6 GHz; and Dual CPU 2.5 GHz PowerPC G5 dual-core. Both an implementation based on logarithmic and exponential look-up tables and a loop based approach are presented. Matrix inversion is performed with a progressive Gauss-Jordan algorithm. After the loop based approach is optimized with SIMD instructions it is shown to significantly outperform the table based approach. To utilize the multi-core architecture of the platforms, the implementation is parallelized. A field size of 2^8 is used and generation sizes between 64 and 256 are tested. Data is coded in blocks of between 128 B and 32 kB. Decoding throughput of up to 43 MB/s is reported at a generation size of 64. The change in decoding throughput obtained with the parallel approach, compared to the single threaded approach[5], ranges from a small decrease at the lowest block size to a speedup of approximately five times at the highest block size.

In [20] an implementation written in Nvidia CUDA is deployed on a PC with an Intel Q6600 2.4 GHz Quad-core CPU and a Nvidia 260 GTX graphics card with 192 cores running at 1.92 GHz. The implementation uses both CPU and GPUs to decode packets. Both single- and multi-threaded matrix inversion techniques are implemented on the CPU. The multiplication operations necessary to decode the data are subsequently performed on the GPU. The combined CPU and GPU approach is used for decoding in order to overcome the problem of synchronizing threads on the GPU. The implementation is based on logarithmic and exponential table look-ups, and matrix inversion is performed with the Gauss-Jordan algorithm[6]. The field size is 2^8 and generation sizes between 128 and 512 are tested. Data is coded in blocks of between 1 kB and 32 kB.

The results from [20] are illustrated in Fig. 4.10[7]. Decoding throughput of up to 225 MB/s is reported at a generation size of 128 and a block size of 32 kB. The coding throughput significantly decreases as the generation size increases and the block size decreases. The matrix inversion on the CPU is observed to be the most important performance bottleneck.

In [21] an implementation written in Nvidia CUDA is deployed on a PC with an Dual Intel Xeon 2.8 GHz Quad-core CPU and a Nvidia 8800 GT graphics card with 112 cores running at 1.5 GHz. The implementation is based on a loop approach using both pure GPU and CPU assisted GPU decoding. The CPU assisted GPU decoding outperforms the pure GPU approach. The authors discuss problems related to pure GPU coding and compared their results to previous CPU based implementation presented in [19]. Matrix inversion is performed with a progressive Gauss-Jordan algorithm. Field sizes of 2^8 are used and generation sizes between 128 and 512 are tested. Data is coded in blocks of size between 128 B and 16 kB.

In [22] the authors extend their work by deploying new optimization techniques on a more powerful Nvidia 280 GTX graphics card with 240

---

[5] Both based on the same SIMD optimized loop based approach.

[6] The matrix inversion approach is not directly stated in the publication but has been confirmed by the authors.

[7] We would like to thank Xiaowen Chu for providing this data.

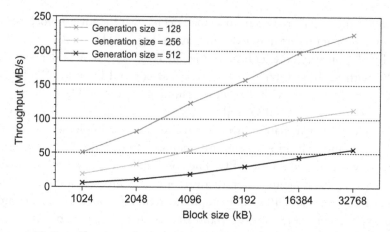

Figure 4.10 Decoding throughput on a powerful desktop PC fitted with a Nvidia 260 GTX graphics card. Generation sizes of 128, 256, 512 are tested along with block sizes between 1 kB and 32 kB [20].

cores running at 1.3 GHz, and compare different implementations based on table and loop approaches. Decoding throughput of up to 254 MB/s is reported at a generation size of 128 and a block size of 32 kB. However, the most interesting contribution of the these publications is the wide range of optimization techniques and different approaches that are compared on mostly uniform hardware. Currently they represent the most thorough treatment of GPU accelerated RLNC implementations. However, as the implementations are based on the Nvidia CUDA architecture they run exclusively on Nvidia hardware.

In [23] an implementation programmed in Nvidia CG and compiled to OpenGL is deployed on a PC with an Intel Q6600 2.4 GHz Quad-core CPU and a Nvidia 9600 GT graphics card with 64 cores running at 1.63 GHz. The implementation is based on table look-ups that are performed on the GPU of the graphics card. Matrix inversion is performed with a progressive Gauss-Jordan algorithm. A field size of 2^8 and generation sizes between 16 and 256 are tested. Data is coded in blocks of 1 kB.

Results obtained are plotted on Fig. 4.11[8]. Decoding throughput of up to 38 MB/s is reported with the generation size of 16. The coding

---

[8] We would like to thank Peter Vingelmann for providing this data.

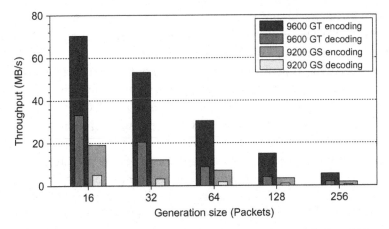

Figure 4.11 Decoding throughput on an OpenGL graphics card. Generation sizes between 16 and 64 are tested at a block size of 1 kB [23].

throughput significantly decreases as the generation size increases and the block size decreases. Unlike previous work, this paper presents an implementation that can be executed on any OpenGL capable platform[9]. This is an intriguing possibility as this could lead to an implementation for mobile phones which support OpenGL and possibly OpenGL ES [24].

8. SPEEDING UP RLNC THROUGH DESIGN

Another approach to improving coding throughput is to change the design of the code used. Here we consider two possible adaptations, namely using a systematic code and coding over the binary field.

In a systematic code all trivial coding vectors[10] are used once, after which encoding vectors are generated randomly. This results in all symbols being sent uncoded once, after which coded symbols are generated and transmitted. This reduces the number of coded symbols that must be generated at the source and decoded at a sink, and thus decreases the computational load on the nodes. The rationale is that all original symbols are

---

[9] In practice it may be necessary to change/port the code to accommodate incompatibilities between different hardware platforms.
[10] In a trivial coding vector all scalars are zero except a single one which is equal to 1.

linearly independent and thus it is not necessary to perform coding until all symbols have been sent once. This approach is simple to implement and reduces the number of expected coded symbols.

When coding is performed over the binary field addition and subtraction is equal to the XOR-operation. Multiplication and division are not needed but are equal to the AND operation. The XOR operation is a very basic operation that all current CPUs can perform quickly, and thus encoding and decoding can be performed much faster compared to when a large field size is used. As the field size decreases the probability that any two randomly generated coded symbols are linearly dependent increases. However, as shown in [25], the probability of linear dependency even with the smallest field possible is surprisingly low. Thus, if the generation size is high enough the loss of degrees of freedom is not significant. Additionally, mobile phones rely almost exclusively on wireless networks, and in such networks the Packet Error Probability (PEP) is typically high and thus receiving a few linearly dependent packets might not be very important compared to a PEP of e.g., 10%. From the energy point of view there is a trade-off between the added cost of additional transmissions versus the reduction in energy consumed at the CPU owing to the lower complexity of the coding operations.

In [25] a Binary RLNC implementation is presented and bench-marked on the Nokia N95-8GB mobile phone, which sports a 332 MHz ARM 11 CPU. All field operations are performed over the binary field and can therefore be performed directly on the CPU. Matrix inversion is performed with a progressive version of the Gauss-Jordan algorithm. The tested field size is 2 and the generation size is between 16 and 256. Data is coded in blocks of 1.5 kB so that it can be transmitted via WLAN without fragmentation.

Results obtained are illustrated in Fig. 4.12. Decoding throughput of up to 15 MB/s for the non-systematic and 29 MB/s for the systematic approach is reported at a generation size of 16. The coding throughput significantly decreases as the generation size increases. It should be noted that the throughput of the systematic approach depends on the ratio between uncoded and coded packets, which depends on the PEP. Hence the non-systematic results are equal to the worst case where all uncoded packets are lost. These results are interesting and may represent an interesting path for implementation of NC, especially for applications

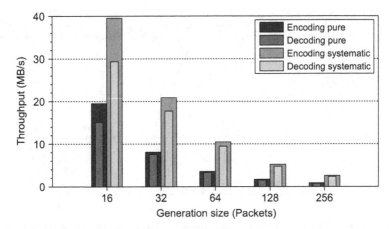

Figure 4.12 Decoding throughput on a Nokia N95-8GB. Generation sizes between 16 and 64 are tested at a block size of 1.5 kB [25].

where a slightly lower network performance is acceptable if it results in reduced energy consumption.

9. A MOBILE PHONE APPLICATION WITH NETWORK CODING

To make a case for the practicality of NC on mobile phones, we visit reference [26] which introduces the PictureViewer application. The PictureViewer application allows users to broadcast images located on their phones to a number of receiving devices. The main idea is that users share content over short range technologies such as WiFi. Instead of uploading the content to social networks such as MySpace or Facebook, the content can be conveyed directly to mobile phones in the vicinity. This application allows all users to see photos on their own mobile devices, which is much more convenient than viewing photos on one device.

The PictureViewer application allows users to monitor the decoding process directly. The decoding process is displayed by drawing the content of the decoding matrix onto the display of the receiving phones. Thus this application provides a very direct feedback which could be useful when suitable protocols are being developed. In Fig. 4.13 the first row of screenshots shows the decoding process of NC. Initially as shown in Fig. 4.13(a) the content of the decoding matrix looks like

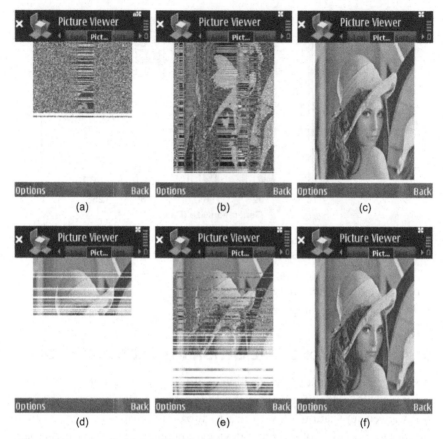

Figure 4.13 Pure NC: (a) Partially decoded data; (b) Image starting to appear as the decoders rank increases; (c) The final decoded image. Systematic NC: (d) Received uncoded data; (e) Erasures corrected by coded packets; (f) The final decoded image.

noise. As the decoder receives more linear combinations and the decoding process starts to solve the decoding matrix, the original picture starts to appear, see Fig. 4.13(b). In the final Fig. 4.13(c) the picture has been decoded and the transmission is complete. In Fig. 4.13 the second row of screen shots shows the decoding process of systematic NC. Figure 4.13(d) shows how uncoded packets are being inserted into the decoding matrix, this continues until all original packets have been sent once. On Fig. 4.13(e) the coding phase has been entered, in this phase erasures which occurred during the uncoded phase are repaired by transmitting encoded packets.

The application is tested on two Nokia N95-8GB mobile phones. One phone acts as the source which transmits a 5 MB file while the other phone is the sink. Tests are performed at different generation sizes and show how the performance of the application is influenced by the increasing complexity of the coding operations. To illustrate the performance impact of the NC operations, the performance of the WLAN when no coding is used is included as a reference. The results without coding indicate the top speed at which the phone can broadcast, and the power consumption during broadcast.

Figure 4.14 illustrates that the throughput decreases as the generation size increases. For low generation size the throughput of coding is similar to that of no coding, which indicates that the computational overhead in this case is not the bottleneck. As the generation size increases the coding operation becomes more computationally demanding and the throughput decreases. Additionally, the systematic approach achieves higher throughput, especially for higher generation sizes. This is expected as the systematic approach increases the coding throughput which is the limiting factor at high generation sizes.

Figure 4.15 shows that the energy consumption increases with the generation size. The reason is that a higher generation size results in

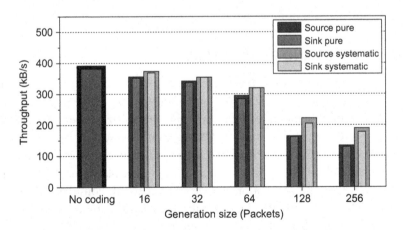

Figure 4.14 Application throughput with and without NC.

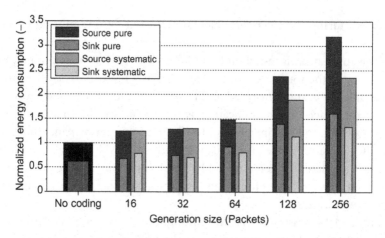

Figure 4.15 Normalized energy consumption of the application with and without NC.

lower throughput and a longer transmission time. The systematic approach decreases this effect by reducing the complexity and thus decreasing the energy consumption. These results give insight into how the use of NC influences the performance of an application. An important finding is that the Nokia N95-8GB mobile phone has such limited computational capabilities that broadcasting data at the rate Wireless Local Area Network (WLAN) supports is problematic. These results show that NC can be deployed even on a low end device like a mobile phone. Future practical experiments will hopefully disclose in which applications and network topologies NC can improve performance.

10. PITFALLS AND PARAMETERS

In an NC system the nodes must agree on at least three parameters, namely the field size, the generation size, and the block size. See Fig. 4.16 for the impact of the choice of field, and generation size. These parameters can be predetermined or included as side information to each encoded piece of data. There exists no general optimal choice of these parameters as they depend on several factors such as the target platform, the network technology, and the application. However, it is worth considering realistic choices when implementing and testing, for example to avoid optimizing towards a setting that will never be used in a real world application. Thus

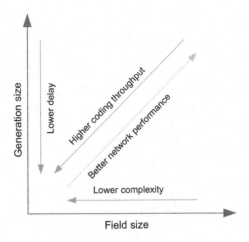

Figure 4.16 Trade-off between low and high field size and generation size.

in the following we present these choices and what requirements different applications might impose.

The field size is an important parameter as it influences the number of unique encoding vectors that can be constructed. In particular, in RLNC the field size determines the probability of receiving a linear dependent combination, which, in turn, determines the performance of the code. The most used field size is currently 2^8 as it provides a reasonable trade-off between code performance and coding throughput. Additionally, this field size is practical if implementation is done with a look-up table based approach because all current platforms implement a data type of this size. Alternatively, a field size of 2 can provide higher coding throughput but also slightly worsens the probability of receiving linearly independent packets. The choice of field size should be based on the target platform and the scenario. For a wireless scenario with a PEP of 10%, a few extra linear dependent packets could be acceptable, as the energy used to perform operations over a large field might be better spent sending a few additional packets. If we instead consider a wired scenario where the PEP is close to 0% and all nodes possess a powerful CPU, the cost[11] of choosing a large

---

[11] In terms of energy and computational resources.

field may be negligible. Hence a good choice of field size depends on the target platform and the scenario in which the system will be deployed.

The generation size also influences the number of unique encoding vectors that can be constructed. As the generation size increases the properties of the code improve, however at the same time the coding throughput decreases, which also influences the energy consumption. In addition the generation size defines the decoding delay, with higher generation size leading to a higher decoding delay. For delay sensitive applications such as VoIP, video conferencing, and gaming, a relatively low generation size must be used. If the delay requirement is strict enough it may not even be beneficial to use NC. In streaming of audio and video services some amount of the media is typically buffered before playback is started, and thus such services are not very sensitive to delay. However, the generation size should be chosen so data in one generation is smaller than the minimum amount of buffered data. For file transfer, P2P, and similar applications, a high generation size can be used, however the influence on coding throughput and energy consumption must still be considered. Thus the choice of generation size should be based on the type of application and the delay requirement.

The block size or packet size can also significantly affect the coding throughput, especially for parallel implementations where a large block size may be necessary to achieve the best throughput. Note that all packets in the encoded block must be received in order for the block to be useful. Note also that if the block size is large, blocks can be fragmented into several frames in the network. If this happens and one frame is lost, the remaining frames will be useless. Hence the block size should be defined based on the network technology over which encoded data is sent, as well as the implementation details.

This chapter focuses on the implementation of NC on commercial mobile platforms. We hope that the chapter will be helpful in understanding several different possible implementations and design choices that must be made when NC is implemented. Furthermore, it outlines some of the results that have been obtained in the field during the last few years, and points towards relevant publications. As a closing remark we underline that it is important to understand the network in which the intended system will be deployed, as well as the requirements of the targeted applications.

REFERENCES

[1] Color Labs, Inc. Color website, April 2011. http://color.com.
[2] P. Vingelmann, F. H. P. Fitzek, and D. E. Lucani. Application-level data dissemination in multi-hop wireless networks. In *IEEE International Conference on Communications (ICC 2010) – CoCoNet Workshop*, May 2010.
[3] D. E. Lucani, F. H. P. Fitzek, M. Médard, and M. Stojanovic. Network coding for data dissemination: It is not what you know, but what your neighbors know. In *RAWNET/WNC3 2009*, June 2009.
[4] R. Jacobsen, K. Jakobsen, P. Ingtoft, T. Madsen, and F. H. P. Fitzek. Practical Evaluation of Partial Network Coding in Wireless Sensor Networks. In *4th International Mobile Multimedia Communications Conference (MobiMedia 2008)*, Oulu, Finland, July 2008. ICTS/ACM.
[5] D. Traskov, J. Lenz, N. Ratnakar, and M. Médard. Asynchronous network coded multicast. In *IEEE International Conference on Communications (ICC 2010)*, May 2010.
[6] P. Sadeghi, R. Shams, and D. Traskov. An optimal adaptive network coding scheme for minimizing decoding delay in broadcast erasure channels. In *EURASIP Journal of Wireless Communications and Networking, Special Issue on Network Coding for Wireless Communications*, pages 1–14, 2010.
[7] Donald E. Knuth. *The art of computer programming, volume 2 (3rd ed.): Seminumerical algorithms*. Addison-Wesley Longman Publishing Co., Inc., Boston, MA, USA, 1997.
[8] Takuji Nishimura and Makoto Matsumoto. Mersenne twister home page. http://www.math.sci.hiroshima-u.ac.jp/~m-mat/MT/emt.html.
[9] Cheng Huang and Lihao Xu. Fast software implementations of finite field operations. Technical report, Department of Computer Science & Engineering, Washington University, St. Louis, MO 63130, December 2003. Available online: http://nisl.wayne.edu/Papers/Tech/GF.pdf.
[10] Neal R. Wagner. *The Laws of Cryptography with Java Code*. 2003. Unpublished, available online: http://www.cs.utsa.edu/~wagner/lawsbookcolor/laws.pdf.
[11] Lieven Vandenberghe. Applied numerical computing. http://www.ee.ucla.edu/~vandenbe/103/reader.pdf.
[12] Morten V. Pedersen, Frank H. P. Fitzek, and Torben Larsen. Implementation and performance evaluation of network coding for cooperative mobile devices. In *IEEE Cognitive and Cooperative Wireless Networks Workshop*. IEEE, May 2008.
[13] S. Katti, H. Rahul, W. Hu, D. Katabi, M. Médard, and J. Crowcroft. XORs in the air: practical wireless network coding. In *Proceedings of the 2006 conference on applications, technologies, architectures, and protocols for computer communications (SIGCOMM '06)*, pages 243–254. ACM Press, September 11–15, 2006.
[14] Mea Wang and Baochun Li. How practical is network coding? *Quality of Service, 2006. IWQoS 2006. 14th IEEE International Workshop on*, pages 274–278, June 2006.
[15] Mea Wang and Baochun Li. Lava: A reality check of network coding in peer-to-peer live streaming. In *INFOCOM*, pages 1082–1090, May 2007.
[16] Szymon Chachulski, Michael Jennings, Sachin Katti, and Dina Katabi. Trading structure for randomness in wireless opportunistic routing. In *SIGCOMM '07: Proceedings of the 2007 conference on applications, technologies, architectures, and protocols for computer communications*, pages 169–180. ACM, August 2007.
[17] Janus Heide, Morten V. Pedersen, Frank H. P. Fitzek, and Torben Larsen. Cautious view on network coding – from theory to practice. *Journal of Communications and Networks (JCN)*, 10(4): 403–411, December 2008.

[18] Hassan Shojania and Baochun Li. Random network coding on the iphone: Fact or fiction? In *NOSSDAV '09: Proceedings of the 18th international workshop on network and operating systems support for digital audio and video*, pages 37–42. ACM, June 2009.

[19] Hassan Shojania and Baochun Li. Parallelized progressive network coding with hardware acceleration. In *Fifteenth IEEE International Workshop on Quality of Service*, pages 47–55, June 2007.

[20] Xiaowen Chu, Kaiyong Zhao, and Mea Wang. Massively parallel network coding on GPUs. In *Performance, Computing and Communications Conference, 2008. IPCCC 2008. IEEE International*, pages 144–151, December 2008.

[21] Hassan Shojania, Baochun Li, and Xin Wang. Nuclei: GPU-accelerated many-core network coding. In *The 28th Conference on Computer Communications (INFOCOM 2009)*, April 2009.

[22] Hassan Shojania and Baochun Li. Pushing the envelope: Extreme network coding on the GPU. In *ICDCS*, pages 490–499, June 2009.

[23] Peter Vingelmann, Peter Zanaty, Frank H. P. Fitzek, and Hassan Charaf. Implementation of random linear network coding on opengl-enabled graphics cards. In *European Wireless 2009*, May 2009.

[24] The Kronos Group. Kronos website, May 2011.

[25] Janus Heide, Morten V. Pedersen, Frank H. P. Fitzek, and Torben Larsen. Network coding for mobile devices – systematic binary random rateless codes. In *The IEEE International Conference on Communications (ICC)*, Dresden, Germany, 14–18 June 2009.

[26] Morten V. Pedersen, Janus Heide, Frank H. P. Fitzek, and Torben Larsen. Pictureviewer – a mobile application using network coding. In *The 15th European Wireless Conference (EW)*, Aalborg, Denmark, 17–20 May 2009.

ABBREVIATIONS FOR CHAPTER 4

CPU	Central Processing Unit
GF	Galois Field
JCN	Journal of Communications and Networks
NC	Network Coding
PEP	Packet Error Probability
RLNC	Random Linear Network Coding
SIMD	Single Instruction, Multiple Data
WLAN	Wireless Local Area Network

Network Coding and User Cooperation for Streaming and Download Services in LTE Networks

Qi Zhang[1], Janus Heide[2], Morten V. Pedersen[2], Frank H.P. Fitzek[2], Jorma Lilleberg[3], and Kari Rikkinen[3]

[1] Aarhus University, Aarhus, Denmark
[2] Department of Electronic Systems, Aalborg University, Aalborg, Denmark
[3] Renesas Mobile Corporation, Oulu, Finland

Contents

Abstract

This chapter proposes the usage of network coding and user cooperation for existing LTE networks. LTE networks are supposed to use Raptor codes for content distribution such as download and streaming services over the air towards mobile devices. Unfortunately, this approach has several shortcomings. Like other FEC technologies, Raptor codes introduce redundancy which increases the total bandwidth of the system, and also increases the energy consumption of mobile devices with good channel characteristics due to ongoing repair messages for devices with bad channel characteristics. Additionally, Raptor codes substantially increase the perceived delay for each user which can be problematic, especially for streaming services. Therefore user cooperation with network coding is proposed to support streaming and download services in future mobile communication systems, overcoming the bandwidth and energy issues described above. The simulation results show that local retransmissions can save up to 80% of redundant information of the cellular link, as long as there are at least two

Network Coding. DOI: 10.1016/B978-0-12-380918-6.00005-6
Copyright © 2012 Elsevier Inc. All rights reserved.

cooperative mobile devices. The results also show that network coding can save more than half of the traffic in the short-range link as long as there are four devices in the cooperation cluster.

Keywords: User cooperation, LTE, Raptor codes, short-range communication.

1. INTRODUCTION

With the increasing demand of diverse bandwidth-demanding services for advanced mobile devices such as video streaming, software distribution, local news, and weather reports, the cellular networks cannot support all the demands by the traditional point-to-point (PTP) transmission methods without requiring prohibitively large amounts of bandwidth. Fortunately, popular services are very often requested by many users. Therefore, a more efficient content delivery method known as point-to-multipoint (PMP) transmission can be used to address this issue. This trend has been recognized by the Third Generation Partnership Project (3GPP) [1] and multimedia broadcast/multicast service (MBMS) has been standardized since 3GPP release 6 [2]. In Release 9 the Evolved MBMS (eMBMS) has been introduced for long-term evolution (LTE) communication systems. Compared with MBMS, eMBMS has improvements in providing high performance using OFDM Multicast/Broadcast Single-Frequency Networking (MBSFN), improving the Signal-to-Interference-plus-Noise-Ratio (SINR) and increasing throughput.

Similar to MBMS, eMBMS supports two types of services for multimedia content delivery, namely *streaming services* and *download services.* The streaming services are used to deliver a continuous data flow of audio and video to mobile devices, while the download services are designed to deliver binary data, still images, text, or software releases. Due to the difference in their characteristics, eMBMS uses different ways to handle streaming and download services. In particular, since streaming services have stringent delay constraints, there is no time for mobile devices to send feedback information such as ACK/NACKs for erasure recovery back to the base station. Therefore, the transmission is considered to be a one-way communication, not involving interactive retransmission. The erasure recovery depends on end-to-end erasure codes, such as Raptor codes [5]. Since no codes can completely guarantee timely and error-free delivery of the data, occasional decoding failures are usually assumed to be tolerable by the end users. On the contrary, download services have no specific requirement on delay, but they require the received data to be error-free in each mobile device. Therefore, it is necessary to retransmit the lost or undecodable packets to guarantee reliable transmission.

To fulfill the needs of these two services, an eMBMS session consists of three phases [3]:

- Initial phase uses two-way PTP communication over TCP/IP or one-way PMP transmission to announce and set up eMBMS services.
- Delivery phase uses one-way PMP transmission supported by protocols using UDP over IP such as RTP (Real-time Transport Protocol) and FLUTE (File Delivery over Unidirectional Transport) to convey streaming or download services.
- Post-delivery phase (optional) uses two-way PTP communication over TCP/IP to report the quality of content reception and to ensure that whole file is delivered to each user.

Figure 5.1 depicts the eMBMS architecture together with its key components. The architecture does not only support the conventional single cell broadcast, but also introduces a new feature, the MBMS single frequency network (MBSFN). The idea of MBSFN is that a number of eNBs (Evolved Node B) broadcast/multicast media content to their UEs

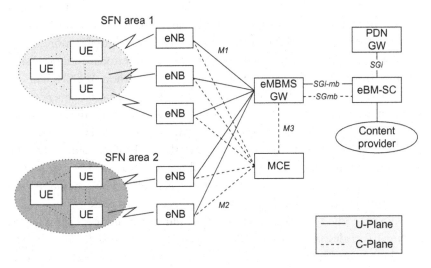

Figure 5.1 The network architecture of eMBMS.

(user equipment) at the same MBMS channel in a time-synchronized mode. MBSFN is promising to deliver the media content to UEs in a more efficient way with improved spectral efficiency. The eMBMS gateway (eMBMS GW) is responsible for distributing the traffic to different eNBs of the MBSFN area. The eMBMS gateway contains user plane and control plane. The eMBMS GW communicates with eNBs for eMBMS data delivery through user plane reference point *M1*. At the same time, eMBMS GW communicates with MCE (Mulitcast Coordination Entity) through control plane reference point *M3*. The MCE ensures that the same resource block is allocated for a given service across all the eNBs of a given MBSFN area. *M2* is the internal control plane interface in E-UTRAN (Evolved UMTS Terrestrial Radio Access Network). Figure 5.1 shows the MCE as a separate entity, but it might be part of the eNB. *SGi-mb* is the reference point between eBM-SC (Evolved Broadcast/Multicast Service Center) and eMBMS GW used for eMBMS data delivery. *SGmb* is the reference point of the control plane between eBM-SC and eMBMS GW. eBM-SC is connected with the packet data network gateway (PDN GW) through the *SGi* interface.

One of the most important challenges in eMBMS is error correction. For both the streaming service and the download service, eMBMS is designed to deliver the content to multiple UEs. The packet erasure encountered by each individual UE could be correlated or independent due to the complicated packet erasure behavior in wireless multicast networks, which will be addressed in detail in Section 3. Furthermore, using the streaming service it is not feasible to use conventional retransmission schemes from eNBs. In the download service, even though eNBs can use the post-delivery phase to repair the missed packets, the post-delivery phase will consume a significant amount of radio resources if a large number of UEs are involved. Therefore, advanced forward error correction technologies have to be employed in eMBMS. In eMBMS, FEC is used at both the physical layer and application layers. At the physical layer, FEC is used to correct bit errors with the 3G air interface channel coding scheme. At the application layer, Raptor codes are used to recover lost packets.

Raptor coding [5] is a new erasure coding technique. Raptor codes extend LT codes [4] by introducing a pre-coding process to improve the decoding probability. Raptor codes are one of the most efficient erasure correction codes currently available. However, Raptor schemes are associated with certain trade-offs between the latency, bandwidth usage, and erasure correction performance.

Erasure correcting codes encode k source symbols into N encoded symbols ($N > k$) so that the source symbols can be reconstructed with a subset of the encoded symbols. To implement Raptor codes in eMBMS, the two parameters, i.e., k and $N - k$, should be considered. Parameter k determines the size of the source block. The size of the source block has an impact on latency and should be small for streaming services. On the other hand, Raptor codes have better performance with large source blocks. Secondly, the number of the additional repair symbols determines the amount of additional bandwidth required to implement FEC. The more repair information, the larger the possibility for the devices to recover lost packets. On the negative side, the need to send a large number of repair symbols consumes a large amount of cellular bandwidth. Furthermore, devices which have received the information correctly need to stay idle until the next transmission phase, resulting in additional energy loss due to the need to monitor the channel in the idle state.

This chapter discusses methods for improving the performance of streaming and download services in LTE networks. Our goal is to minimize the latency of the services and decrease the bandwidth usage by employing user cooperation and network coding. In order to be standard compliant to the 3GPP specifications, we keep the Raptor coding in the cellular link. User cooperation leverages the potential to establish local links between closely located mobile devices using short-range communication technologies such as Bluetooth and WiFi in addition to their cellular links. In this setting mobile devices are not acting alone but form a cooperative group, sometimes referred to as a *wireless grid*. Cooperative clusters have several benefits over standalone devices in terms of spectral efficiency, energy consumption, and achievable data rates (see [6, 7, 9–11]). In this chapter user cooperation is used for local retransmissions within the cooperative cluster to reduce the load on cellular links. The exchange of packets within the cluster is facilitated by using the network coding technique in order to keep the overall necessary transmissions in the cluster low.

The chapter is organized as follows: Section 2 describes the basic theory of Raptor codes and presents the trade-offs associated with Raptor codes implementation in eMBMS. Section 3 explains the packet erasure pattern in the wireless multicast scenario. User cooperation for packet erasure recovery will be proposed and described in Section 4. Section 5 explains how to apply network coding to the local retransmission with user cooperation to improve the local retransmission efficiency. Simulation results will be presented in Section 6. The chapter is concluded in Section 7.

2. RAPTOR CODES IN eMBMS

Raptor codes provide packet-level protection at the application layer to complement the bit-level FEC at the physical layer. Raptor codes are implemented on both the evolved Broadcast Multicast Service Center (eBMSC) and on each individual mobile device. The basic encoding process is as follows: a complete data file or a segment of a data stream is inserted into a large data block which is referred to as the *source block*. One source block has k symbols and each symbol is composed of T bytes. Hence, one source block has kT bytes. The structure of the source block is shown in Fig. 5.2.

After constructing a completed source block, the Raptor encoder generates $N - k$ repair symbols, each of size T. The number of repair symbols, i.e., $N - k$, depends on whether the source block is used for eMBMS streaming service or download service, the anticipated network conditions, the desired quality of service, the amount of available additional bandwidth, or the allowed transmission time [3]. The eBMSC uses a *systematic code*, i.e., it sends k source symbols followed by $N - k$ repair symbols to all the receivers.

The advantage of a systematic code is that some of the received source symbols do not require decoding, hence they can be used by the receiver directly. Even in the case where a source block cannot be decoded, the original data in the first k symbols can be passed on to the higher layers,

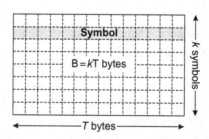

Figure 5.2 The structure of the source block.

which may cope with the errors by error concealment algorithms. At the receiver side, as long as the receiver collects enough packets (source or repair packets), the receiver is able to decode the source block. If, due to multiple packet erasures, the receiver does not receive enough packets to decode a source block, only the part of the source block that is decoded can be delivered to higher layers.

The general framework of streaming and download services in eMBMS is illustrated in Fig. 5.3. Both streaming and download services use a similar framework. The main difference is that steaming media applications use RTP (Real-time Transport Protocol) to transfer the data while non real-time applications use FLUTE (File Delivery over Unidirectional Transport Protocol) to transport data. A detailed description of RTP and FLUTE can be found in [3].

To choose the right source block size k and the number of repair symbols $N - k$, the trade-offs between source block size, latency, expected quality of service, repair symbols, and bandwidth expansion have to be taken into consideration.

Let us consider the streaming service as an example. For the streaming service, the *mean time between failures* (MTBF) is often used as a metric to

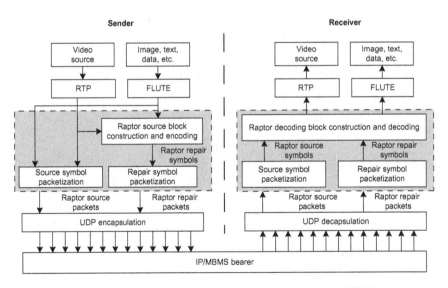

Figure 5.3 The framework of streaming and download services in eMBMS.

measure the expected Quality of Service (QoS). Suppose that each source block contains t_s seconds of video content, and that the probability of decoding failure is P_{df}. The expected time between decoding failure is then t_s/P_{df} seconds.

To meet the MTBF requirement τ, it must hold that:

$$\frac{t_s}{P_{df}} \geq \tau. \tag{1}$$

The probability of decoding failure of Raptor codes can be modeled as follows [12]:

$$P_{df} = \begin{cases} 1 & \text{if } m < k, \\ 0.85 \times 0.567^{m-k} & \text{if } m \geq k, \end{cases} \tag{2}$$

where P_{df} denotes the decoding failure probability of the code with k source symbols if m symbols have been received [12].

For a given anticipated packet error rate p_e, assuming the eBMSC sends N symbols in total for each source block and each symbol is encapsulated into one packet, then the expected number of the successfully received symbols can be expressed by:

$$m = N(1 - p_e). \tag{3}$$

By Equations (1), (2), and (3), to satisfy the reliability requirement, the number of symbols at each block must be at least:

$$N \geq \frac{1}{1 - p_e} \left(\log_{0.567} \frac{t_s}{0.85\tau} + k \right)$$

$$\geq \frac{1}{1 - p_e} \left(k - 1.7624 \ln \frac{t_s}{\tau} - 0.2864 \right). \tag{4}$$

Therefore, the minimum number of repair packets for each block, R, is:

$$R = N - k \tag{5}$$

$$= \frac{1}{1 - p_e} \left(p_e \cdot k - 1.7624 \ln \frac{t_s}{\tau} - 0.2864 \right).$$

If the data rate of video streaming is r bit/s, then the perceived delay D can be expressed by:

$$D = t_s = \frac{8kT}{r}. \tag{6}$$

According to Equation (6), it is not difficult to see that larger source block sizes result in longer perceived delays. For example, suppose that the targeted MTBF is one hour, the video streaming playout data rate is 512 kb/s, the symbol size T is 256 bytes, and the average packet error rate is 10%. For this setting, Table 5.1 lists the overhead and delay for different source block sizes. The overhead here is defined as the number of redundant packets divided by the number of information packets, i.e., R/k. From the table, we can see that the delay is increasing linearly with an increasing source block size. In contrast, the overhead decreases with the increase of the source block size. Thus, to satisfy the perceived delay constraints, small block sizes are preferred. On the other hand, more repair symbols are required with small block sizes than with larger block sizes to achieve the same performance.

Furthermore, the overhead presented in Table 5.1 is the ideal theoretical value that assumes that the packet erasure rate is known. In reality, the packet erasure pattern of a wireless multicast network is affected by various factors such as channel conditions and the quality of the receiver's hardware, and it is difficult to predict (see Section 3 for more discussion). The average packet error rate can vary within a block and can be different for different blocks, especially when the block size is small. In the next section, the impact of packet erasure patterns on Raptor code performance will be further addressed.

The example above is for streaming service. For download service the overhead is even more significant, because download service requires error free reception. The missed symbols need to be retransmitted from the eNB in the post-delivery phase, which will cost additional time and bandwidth.

Table 5.1 Overhead and delay for different source block length

Source block length	1024	2048	8192
Overhead	0.1238	0.1168	0.1122
Delay (s)	4.0960	8.1920	32.7680

3. PACKET ERASURE PATTERN

The packet erasure pattern among mobile devices in wireless multicast networks varies dramatically. The reason is that packet erasure can occur due to several reasons, for instance, network congestion, deep fading, severe path loss, interference, and hardware performance. Hardware performance is often underestimated in many theoretical research works.

Packet erasures in multicast wireless networks can be correlated and (or) uncorrelated. The packet erasures within the *coherence time*[1] are often correlated. However, the majority of packet erasures at mobile devices are independent, and each mobile device has heterogeneous packet erasure behavior. The reasons are mainly due to the different locations of nodes and to manufacturing variability. In eMBMS, the mobile devices are located at different locations in the same cell or even different cells, which results in different path loss, shadowing, received interference, etc. at each receiver. Therefore, the channel conditions seen by mobile devices can be very different from each other. Furthermore, many real measurement results [13–15] reveal the interesting fact that, even though the mobile devices have the same software and hardware, there is a large variation in their performance. This is due to manufacturing differences which can be significant. In [14] network layer measurements using Nokia N95 mobile phones showed significant variability in the packet error probability, with some devices showing high packet error probability during all tested setups. Those findings were verified in [13] where the frame error rate varied significantly for different cards using the RT2500 chipset under the same measurement conditions.

In a nutshell, the packet erasure pattern among all the mobile devices in eMBMS is a combination of correlated and independent, random and burst erasures.

In Section 2, it was mentioned that Raptor codes have better performance with larger source block size. One of the reasons is that with a large block size the number of lost packets over each block will have a higher probability of approaching its average value according to the law of large numbers. It means that the number of repair symbols generated by the

---

[1] Coherence time is the minimum time required for the magnitude change of the channel to become uncorrelated from its previous value.

Raptor code is close to the predicted value and it has a larger probability to meet the expected QoS. However, for small block sizes the number of lost packets over each block can vary over a wide range. This means that the number of generated repair packets in a Raptor code might not be enough to achieve the desired level of QoS.

Figure 5.4 shows a cumulative distribution function of the actual packet error rate for different source block length cases. Assuming the average packet error rate is 10%, we simulate transmission of 1000 blocks with a source block lengths (SBL) of 1024, 2048, and 8192. Our results indicate that 99% of the blocks have a packet error rate of less than 12.1%, 11.4% and 10.6%, respectively. Therefore, to be able to decode 99% of the blocks, in the case of SBL = 1024, the encoding process of a Raptor code needs to generate repair symbols for a packet error rate equal to 12.1%. In the case of SBL = 8192, it will be sufficient to generate repair symbols at a packet

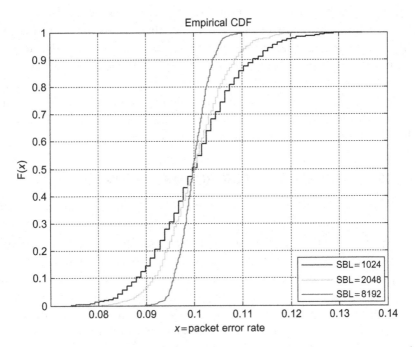

Figure 5.4 Cumulative Distribution Function of the actual packet error rate in each block with a different source block length.

Table 5.2 Theoretical and practical overhead comparison with different source block length

Source block length	1024	2048	8192
Overhead (Theoretical)	0.1238	0.1168	0.1122
Overhead (Practical – 1 node)	0.1493	0.1344	0.1197
Overhead (Practical – 4 nodes)	0.1585	0.1396	0.1241

error rate of 10.6%. Therefore, the overhead introduced by repair symbols in practice is larger than the theoretical value.

When there are two or more mobile devices in the multicast group, the system has to use the packet error rate of the worst mobile device. For instance, if there are four mobile devices receiving the multicast content in the case of SBL = 1024, 2048, and 8192 respectively, the actual worst case packet error rates among the four nodes are 12.7%, 11.8%, and 10.95%. Table 5.2 compares the results obtained with the theoretical values. The values in the table show that the overhead for small block sizes is larger than that of large block sizes. In this example, the overhead of SBL = 1024 is 24.7% more than that of SBL = 8192 in the single-node case. Also, it shows that the overhead of the four-node case is larger than that of the single-node case. It indicates that, when the number of mobile devices increases, the overhead is likely to increase. It is worth mentioning that a source block size of 1024 is the minimum source block size defined in 3GPP [16]. The main reason is that when the source block size is smaller than 1024, the overhead is so big that it affects the efficiency.

4. USER COOPERATION FOR ERASURE RECOVERY

The concept of user cooperation was proposed in [6, 17]. The basic idea of user cooperation is that a mobile device can exchange packets with the neighboring devices in its proximity over short-range links to achieve a common or individual goal. The generic network architecture of user cooperation is shown in Fig. 5.5. Since most off-the-shelf mobile devices are all multi-mode devices, namely they not only have cellular interface, but also short-range link interface, such as WiFi, there is an opportunity to implement user collaboration schemes in such devices. It has been proven to be a promising scheme to tackle many challenges that the cellular network itself cannot address. Many research works, such as [9, 11, 18, 19], have

R
a   C
p   o
t   d
o   e
r   s

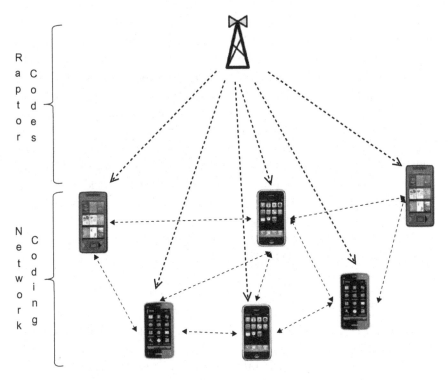

N
e   C
t   o
w   d
o   i
r   n
k   g

Figure 5.5 Network architecture of user cooperation.

shown the gain of user cooperation in energy saving, throughput enhance-
ment, etc. In this chapter, user cooperation is applied for erasure recovery in
eMBMS. We use user collaboration to reduce the overhead introduced by
repair symbols. Furthermore, with user collaboration, smaller source blocks
can be used without degrading the error correction performance of Raptor
codes.

As we mentioned in Section 3, even though the packet erasure pattern is a combination of many factors in multicast wireless network, the
uncorrelated packet erasure among mobile devices is dominant in all the
lost packets. This means that the neighboring devices often have the packets that the mobile device has lost. In other words, those devices can use
user cooperation to help each other to recover the lost packets.

For eMBMS with Raptor codes, the proposed cooperative erasure
recovery scheme works as follows. As the sender sends out a batch of coded

packets for each block, it takes a few seconds for the sender to finish sending out all the packets for each block. Likewise, it will take the receiver a few seconds to receive enough packets to construct the decoding block. The encoding and decoding procedure of Raptor codes is detailed in 3GPP TS26.346 [8]. In the proposed cooperative erasure recovery scheme, it is not necessary for the mobile devices to start erasure recovery until the end of a block. The mobile devices can progressively recover the erasures locally. For instance, local recovery can be performed every 64 packets, which not only reduces the number of the needed repair symbols from eNB, but also allows a large amount of repair symbols to be obtained quickly, which, in turn, reduces the decoding time at the end nodes. The progressive recovery procedure is demonstrated in Fig. 5.6.

In local retransmission with user cooperation, the set of the all missed packets of a cluster can be expressed by:

$$\hat{L} = L_1 \cup L_2 \ldots \cup L_n, \tag{7}$$

where $L_1, L_2, \ldots, L_n$ are the sets of the missed packets of mobile devices $1, 2, \ldots, n$, respectively.

The common lost packets in the cluster can be written as:

$$\Lambda = L_1 \cap L_2 \ldots \cap L_n. \tag{8}$$

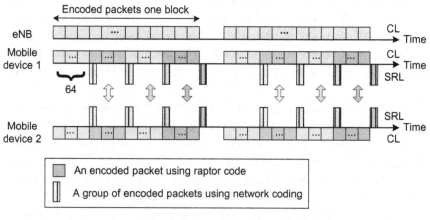

Figure 5.6 Illustration of progressive recovery in user cooperation.

When $\Lambda = \phi$, all the missed packets can be recovered within the cluster. The total number of the recoverable packets by local retransmission, N_{rv}, is given by:

$$N_{rv} = |\hat{L}| - |\Lambda|, \tag{9}$$

where $|\hat{L}|$ is the cardinality of set $\hat{L}$, i.e., the number of the total lost packets among the mobile devices, and $|\Lambda|$ is the number of packets lost by all devices in the cluster.

With local retransmissions, eBMSC can send N_{coop} symbols for each source block, which includes source symbols and repair symbols. Then the number of the successfully received symbols, m_{coop}, can be expressed as:

$$m_{coop} = N_{coop} - |\Lambda|. \tag{10}$$

According to Equations (1), (2), and (10), the minimum number of symbols that should be sent by eBMSC can be obtained from the following equation:

$$N_{coop} \geq k + |\Lambda| - 1.7624 \ln \frac{t_s}{\tau} - 0.2864. \tag{11}$$

With user cooperation, the minimum number of repair symbols for each source block, R_{coop}, becomes:

$$R_{coop} = N_{coop} - k \tag{12}$$

$$= |\Lambda| - 1.7624 \ln \frac{t_s}{\tau} - 0.2864.$$

In the cellular link, user cooperation based local retransmission can reduce the minimum number of required repair symbols for each block from R to R_{coop}. Therefore, the user cooperation gain, G_{coop}, can be written as follows:

$$G_{coop} = \frac{R - R_{coop}}{R} \tag{13}$$

$$= 1 - \frac{(1 - p_e^c)(|\Lambda| + \xi)}{p_e^c \cdot k + \xi}$$

where:

$$\xi = -1.7624 \ln \frac{t_s}{\tau} - 0.2864$$

and p_e^c is the cellular link packet erasure rate of the worst mobile device in the multicast group.

Comparing the cooperation case with the case of no user cooperation, we may see from Equation (13) that the cooperation gain depends on $|\Lambda|$, the number of packets not delivered to any mobile device. The smaller $|\Lambda|$ is, the larger the cooperation gain will be. For example, when $k = 1024$, $T = 256$, playout speed 512 kbps, the MTBF (mean time before failures) 3600 seconds, packet error rate $p_e^c = 10\%$, $|\Lambda|$ is 10% of the total erasures $|\hat{L}|$, and the cooperations gain is 82.72%. Cooperation gain, G_{coop}, reaches its upper bound when $|\Lambda|$ equals to zero. In this case, the maximum cooperation gain is 90.8% (see Table 5.3). Note that 90.8% is a theoretical value based on the assumption that the packet erasure rate is equal to the average packet erasure rate.

Furthermore, the local retransmission with user cooperation can save a large amount of the retransmission traffic in the post-delivery phase for download service, as most of the erasures in the cluster can be recovered over the short-range links. Additionally, it takes less time for each mobile device to recover all the erasures, as the data rate of a short-range link is usually higher than that of the cellular link.

The cooperation gain achieved by using short-range links is very impressive; however, it has some costs associated with using short-range links. The cost includes two parts: the communication cost of mobile devices, and the network resource cost.

The communication cost of mobile devices is mainly the energy used in short-range communication. The energy per bit ratio (EpBR) of a short-range link is much lower than that of a cellular link. Thus, the additional energy cost at the cellular link is small compared to the energy saved in the cellular link. In other words, there is a substantial energy saving gain in the

Table 5.3 Cooperation gain

$|\Lambda|$	0	$0.1|\hat{L}|$	$0.2|\hat{L}|$	$0.3|\hat{L}|$	$0.4|\hat{L}|$	$0.5|\hat{L}|$
G_{coop}	0.9080	0.8272	0.7464	0.6656	0.5848	0.5040

overall energy consumption of the mobile devices even though user cooperation incurs additional energy cost at the short-range link. For detailed discussion on leveraging user cooperation to reduce energy consumption in wireless networks see references [9, 11, 18, 19].

To evaluate the network resource cost, it is necessary to calculate the number of exchanged packets over the short-range link, N_{sr}. This can be expressed as:

$$N_{sr} = \frac{N_{rv}}{1 - p_e^{sr}}$$

$$= \frac{|\hat{L}| - |\Lambda|}{1 - p_e^{sr}},$$

(14)

where p_e^{sr} is the packet error rate in the short-range link. Note that in practice, p_e^{sr} is smaller than p_e^{c}.

On the one hand, as short-range links usually use the free spectrum and have a much higher data rate than that of the cellular link, the network resources used for packet exchange over a short-range link are usually regarded as free. On the other hand, considering more and more applications are starting to exploit the lower cost short-range links, there is a need to use the short-range network resources in an efficient way. The next section will address how to exchange packets over the short-range link in a more efficient way.

5. NETWORK CODING APPLIED IN USER COOPERATION

The main motivation of applying network coding to user cooperation is to improve the cooperation efficiency for the short-range links. First of all, network coding can significantly reduce the number of packets exchanged among mobile devices. In particular, with network coding the number of exchanged packets required to decode all the recoverable erasures in the cooperative cluster could be less than $|\hat{L}| - |\Lambda|$. As an encoded packet is a combination of multiple packets, many nodes that lost different packets can benefit from the same encoded packet. Furthermore, considering partially connected cooperative clusters, i.e., the clusters where peers cannot communicate directly with each other, but information can be relayed within the cluster, the recoding characteristics of network coding can make the packet exchange very efficient. Therefore, network coding can help to reduce the overall traffic and energy used for cooperative exchange over the short-range link. In the following, we analyze the number of coded packets exchanged locally.

Let i and j be the mobile devices that have the highest and the lowest number of lost packets, respectively, i.e.:

$$|L_i| = \max\{|L_1|, |L_2|, \ldots, |L_n|\} \tag{15}$$

$$|L_j| = \min\{|L_1|, |L_2|, \ldots, |L_n|\}. \tag{16}$$

In the case where mobile devices i and j do not have any common erasures, mobile device j can send coded packets to repair all the erasures in mobile device i, and *vice versa*. These coded packets can also be used to correct the erasures at other mobile devices in the cooperation cluster. Even though they have different erasures, as long as the number of the recoverable erasures of the other mobile devices is less than that of mobile device i, it is possible to construct an encoding scheme that ensures that all the other mobile devices can recover their erasures by overhearing the coded packets.

When mobile device i gets its erasures corrected, it can send coded packets to correct the erasures of mobile device j. In this case, the number of the exchanged coded packets in the cooperative cluster, N_{nc}, can be expressed as:

$$N_{nc} = (|L_i| - |\Lambda|) + (|L_j| - |\Lambda|) \tag{17}$$

where $|\Lambda|$ is the number of lost packets of the cluster, i.e., the correlated lost packets of the mobile devices which are not recoverable by local retransmission.

In the case that mobile device i and j have some common erasures besides the ones in Λ, the rest of the nodes can help to correct these erasures. In this case N_{nc} is less than that of the former case, i.e.:

$$N_{nc} < (|L_i| - |\Lambda|) + (|L_j| - |\Lambda|), \tag{18}$$

which gives an upper bound of the exchanged coded packet in the short-range link.

Next, we look at the issue from another angle. The set of packets only received by mobile device k is denoted by Δ_k. Then it holds that:

$$\Delta_k = \Lambda_{-k} \setminus \Lambda, \qquad (19)$$

where $\Lambda_{-k} = L_1 \cap L_2 \cap \ldots L_{k-1} \cap L_{k+1} \cap \ldots \cap L_n$.

To decode all the recoverable erasures at mobile device k it must receive $|L_k| - |\Lambda|$ packets. Mobile device k must also send $|\Delta_k|$ packets, as it is the only one that holds these packets. Therefore, the number of exchanged packets that node k sends and receives is expressed by $|\Omega_k|$:

$$|\Omega_k| = (|L_k| - |\Lambda|) + |\Delta_k|. \qquad (20)$$

Let us assume $|\Omega_k|$ is the largest among $\{|\Omega_1|, |\Omega_2|, ..., |\Omega_n|\}$. If $|L_i| \leq \max\{|\Omega_1|, |\Omega_2|, ..., |\Omega_n|\}$, then after mobile device k has exchanged $|\Omega_k|$ coded packets by network coding in the cluster, this procedure does not only help mobile device i to recover the missed packets, but also distributes the unique $|\Delta_k|$ packets of mobile device k among the cluster. Therefore:

$$N_{nc} = \max\{|\Omega_1|, |\Omega_2|, ..., |\Omega_n|\}. \qquad (21)$$

If $|L_i| > \max\{|\Omega_1|, |\Omega_2|, ..., |\Omega_n|\}$, then the exchanged coded packets among the cluster must be at least $|L_i|$. Therefore:

$$N_{nc} > \max\{|\Omega_1|, |\Omega_2|, ..., |\Omega_n|\}, \qquad (22)$$

which gives a lower bound on the number of exchanged packets in the short-range link.

Thus, we obtain the range of N_{nc} as:

$$\max\{|\Omega_1|, |\Omega_2|, ..., |\Omega_n|\} \leq N_{nc} \leq |L_i| + |L_j| - 2|\Lambda|. \qquad (23)$$

N_{nc} is the ideal number of exchanged packets in a short-range link. To be precise, the linearly independent probability of the received coded packets and the packet error rate in the short-range link should be taken into account.

Given that the sender holds g linearly independent symbols, the probability that a received coded symbol is linearly independent is given in [20] by:

$$P_{ind} = 1 - \frac{1}{q^{g-g'}}, \qquad (24)$$

where g' is the number of the received independent symbols at the sink. Thus the expected number of coded symbols that must be received to decode a generation can be calculated as:

$$E_g = \sum_{g'=0}^{g-1} \left(1 - \frac{1}{q^{g-g'}}\right)^{-1}. \qquad (25)$$

Hence, the number of the exchanged coded packets to correct all the recoverable erasures locally is expressed by:

$$N_{sr}^{nc} = \frac{N_{nc} \cdot E_g}{1 - p_e^{sr}}. \qquad (26)$$

When q is high, $E_g \approx 1$, thus:

$$N_{sr}^{nc} \approx \frac{N_{nc}}{1 - p_e^{sr}}. \qquad (27)$$

The value of N_{nc} in practice depends on the local retransmission scheme with network coding. In other words, if the local retransmission scheme is designed well, N_{nc} will be close to the lower bound. There are many possible schemes to implement local retransmission. We propose one as an example. The basic idea is that the *current* "best" mobile device, i.e., the one with the least packet erasures, first sends an encoded packet with all the packets it has received.

The benefit of such an encoded packet is two-fold. On the one hand, the encoded packet has the highest probability of correcting the most erasures in the other mobile devices. On the other hand, it implicitly indicates which packets it misses. The missed packets of the *current* "best" mobile device are regarded as the *current* "rare" packets. Before a mobile device sends the coded packets, it will wait a certain time period according to its back-off timer. The value of the timer is a function of the number of packets and the number of rare packets the mobile device has. The more "rare" packets a mobile device has, the lower the back-off time is. Thus by receiving the coded packets, some "non-best" mobile devices become "better", namely the mobile devices have less erasures.

Then one of these mobile devices will become the *current* "best" mobile device. The *current* "best" mobile device will start sending coded packets until it is replaced by another "better" one. As soon as the *current* "best" mobile device sends out an encoded packet which includes all the packets of this generation, all of the others reset their back-off timer. Then the back-off timer is used for sending a feedback, the value of which is a function of the number of packets a mobile device needs. The more packets a mobile device still needs, the less the back-off time is. Thus the *current* "worst" mobile device can give a short feedback to indicate how many packets it still needs. Then the *current* "best" mobile device will stop sending after sending out the required number of packets.

6. SIMULATION RESULTS

In this section the two main simulation results are presented. First, the gain of local retransmission with user cooperation and the gain of network coding applied to the short-range link are given.

According to the analysis in Section 2, we know that the erasure correction performance depends on the source block size. The smaller the source block size is the more overhead it carries. 3GPP limits the minimum source block size to 1024 due to the inefficiency of having a block size lower than 1024, even though lower values would be beneficial for ensuring the quality of service. With user cooperation it can further reduce the source block size to 512 or even 256. Smaller block sizes will reduce the perceived latency.

Let us assume that the average packet error rate is 10%. The simulation results of user cooperation are shown in Fig. 5.7. When there are two, three, and four mobile devices in the cooperation cluster, and the block size is 1024, it can save 80.6%, 89.0%, and 92.4% overhead in the cellular link, respectively. It clearly shows the significant overhead saving by user cooperation. It also shows that Raptor coding becomes nearly obsolete for

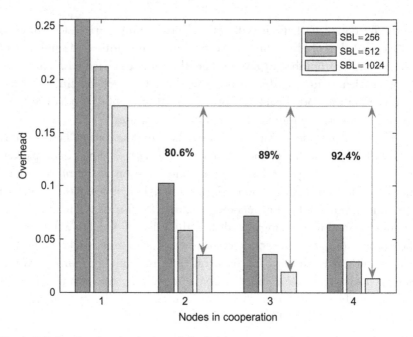

Figure 5.7 Overhead saving in the cellular link by user cooperation.

large cooperation clusters. Furthermore, as long as there are two devices in the cooperation cluster, the overhead of block size equal to 512 and 256 is only 6% and 10%, respectively. Such overhead is lower than the overhead in the case that block size is equal to 1024 and no cooperation is involved. It means that it is feasible to use a smaller block size such as 512 and 256 with two users. It is also clear that using block size 512 and 256, the perceived delay can be reduced by 50% and 75%, respectively. To sum up, in the proposed state-of-the-art system, the overhead can only be reduced by sacrificing latency and *vice versa*. User cooperation on the other side offers both low latency and low overhead.

To show the main benefit of network coding, we compare the number of the exchanged packets to recover all the erasures in normal user cooperation and user cooperation with network coding. The basic assumption of the simulation is that the local retransmission is done every 64 packets, i.e., the generation size of network coding is 64. The packet error rate of cellular link and short-range link is 10% and 5%, respectively. The simulation result are shown in Figs. 5.8 and 5.9. Network coding starts to work when

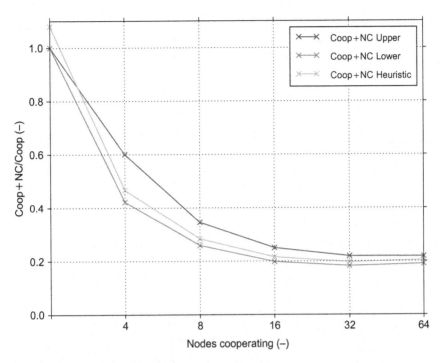

Figure 5.8 Cooperation cost saving by network coding over the short-range link.

there are more than two cooperation devices. It shows that cooperation with network coding needs much less packet exchange than cooperation without network coding. In other words, cooperation with network coding saves the cooperation cost over the short-range link in terms of the number of exchanged packets. We can see from Fig. 5.8 that, for instance, when there are four devices in the cooperative cluster, network coding heuristic approach saves more than half of the exchanged packets; it can save about 75% of exchanged packets when there are eight devices. Furthermore, we can tell that cooperation with network coding can recover all the erasures in a shorter time than cooperation without network coding, as less packet exchange is needed. Figure 5.9 shows that the number of exchanged packets increases smoothly with network coding with an increasing number of cooperating nodes. However, it increases dramatically without network coding. This means that cooperation with network coding has better scalability in a big cooperation cluster. Additionally, both Figs. 5.8 and 5.9 show

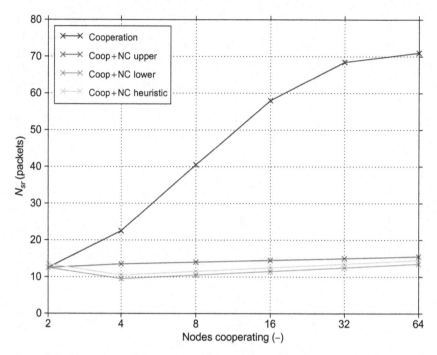

Figure 5.9 Comparison of the number of the exchanged packets over the short-range link with/without network coding.

that the performance of the proposed local retransmission scheme with network coding, i.e., the heuristic network coding approach, is very close to the derived lower bound.

7. CONCLUSION

To tackle the drawbacks of Raptor code in eMBMS, a novel local retransmission scheme based on the concept of user cooperation has been proposed in addition to the usage of Raptor codes. The simulation results show that local retransmission can save about 80% overhead in the cellular link as long as two or more mobile devices cooperate. Larger gains can be achieved by increasing the number of cooperating devices. Furthermore, local retransmission makes it feasible to use smaller block sizes on the cellular link using Raptor codes to reduce the user perceived delay and to improve user perceived QoS. To make the local retransmission in the short-range link more efficient, network coding is considered for the local retransmission and a local retransmission protocol is proposed. The simulation results show that network coding can save more than half of the short-range link traffic as long as there are four mobile devices in the cooperation cluster. Reducing the traffic on the short-range link can reduce the overall energy consumption, as well as reduce the time that is needed to complete the exchange of packets locally, especially in dense traffic networks.

REFERENCES

[1] http://www.3gpp.org/
[2] 3GPP TS 26.346 V6.6.0, Technical Specification Group Services and System Aspects; Multimedia Broadcast/Multicast Service (MBMS); Protocols and codecs, Oct. 2006.
[3] How DF Raptor is Used in MBMS. Technical report, Digital Fountain, March 2007.
[4] M. Luby. LT Codes. In *43rd Annual IEEE Symposium on Foundations of Computer Science (FOCS)*, pages 271–282, Vancouver, Canada, Nov 2002.
[5] A. Shokrollahi. Raptor codes. *IEEE Transactions on Information Theory*, 52(6):2551–2567, June 2006.
[6] Frank H.P. Fitzek and Marcos D. Katz, editors. *Cooperation in Wireless Networks: Principle and Applications*. ISBN-10 1-4020-4710-X. Springer, 2006.
[7] Frank H.P. Fitzek and Katz Marcos, editors. *Cognitive Wireless Networks: Concepts, Methodologies and Visions*. ISBN: 978-1-4020-5978-0. Springer, 2007.
[8] 3GPP TS 26.346 V8.1.0, *Technical Specification Group Radio Access Network; Multimedia Broadcast/Multicast Service; Protocols and Codec*, Dec 2008.
[9] Q. Zhang, F.H.P. Fitzek, and Marcos Katz. Cooperative Power Saving Strategies for IP-Services Supported over DVB-H Networks. In *Wireless Communication and Network Conference*, March 2007.
[10] Qi Zhang, Frank H.P. Fitzek, and Villy B. Iversen. Design and Performance Evaluation of Cooperative Retransmission Scheme for Reliable Multicast Services in Cellular Controlled P2P Networks. In *The 18th Annual IEEE International Symposium on Personal, Indoor and Mobile Radio Communications (PIMRC07)*, September 2007.
[11] Qi Zhang, Frank H.P. Fitzek, and V. B. Iversen. One4all Cooperative Media Access Strategy in Infrastructure Based Distributed Wireless Networks. In *IEEE Wireless Communication and Networking Conference (WCNC)*, April 2008.
[12] M. Luby, M. Watson, T. Gasiba, and T. Stockhammer. Mobile Data Broadcasting over MBMS Tradeoffs in Forward Error Correction. In *the 5th International Conference on Mobile and Ubiquitous Multimedia (MUM)*, Stanford, CA, USA, December 2006.
[13] P. Fuxjager and F. Ricciato. Collecting Broken Frames: Error Statistics in IEEE 802.11b/g Links. In *Modeling and Optimization in Mobile, Ad Hoc, and Wireless Networks and Workshops, 2008. WiOPT 2008. 6th International Symposium on*, pages 30–35, April 2008.
[14] Janus Heide, Morten V. Pedersen, Frank H.P. Fitzek, Tatiana V. Kozlova, and Torben Larsen. Know Your Neighbour: Packet Loss Correlation In IEEE 802.11b/g Multicast. In *4th International Mobile Multimedia Communications Conference (MobiMedia)*, Oulu, Finland, 7–9 July 2008.

[15] Hoi-Sheung Wilson So, Kevin Fall, and Jean Walrand. Packet Loss Behavior in a Wireless Broadcast Sensor Network. http://citeseerx.ist.psu.edu/

[16] M. Luby, A. Shokrollahi, M. Watson, and T. Stockhammer. Raptor Forward Error Correction Scheme for Object Delivery, RFC 5053, October 2007.

[17] F.H.P. Fitzek, M. Katz, and Q. Zhang. Cellular Controlled Short-Range Communication for Cooperative P2P Networking. In *Wireless World Research Forum (WWRF) 17*, volume WG 5, Heidelberg, Germany, November 2006. WWRF.

[18] F.H.P. Fitzek, M. Katz, and Q. Zhang. Cellular Controlled Short-range Communication for Cooperative P2P networking. *Wireless Personal Communications*, 2008.

[19] T.K. Madsen, Q. Zhang, and F.H.P. Fitzek. Design and Evaluation of IP Header Compression for Cellular-Controlled P2P Networks. In *IEEE International Conference on Communication (ICC)*, June 2007.

[20] A. Eryilmaz, A. Ozdaglar, and M. Medard. On Delay Performance Gains From Network Coding. *Information Sciences and Systems, 2006 40th Annual Conference on*, pages 864–870, March 2006.

CHAPTER 6

CONCERTO: Experiences with a Real-World MANET System Based on Network Coding

Victor Firoiu[1], Greg Lauer[2,*], Brian DeCleene[1], and Soumendra Nanda[1]

[1]Advanced Information Technologies, BAE Systems, Burlington, MA, USA;
[2]BBN Technologies, Cambridge, MA, USA

Contents

Distribution Statement A: Approved for Public Release; Distribution Unlimited. Cleared for Open Publication on 2/24/2009, 8/18/2009 and 5/10/2010.

This work was funded by DARPA under contract N66001-06-C-2020.

The views expressed herein are those of the authors and do not reflect the official policy or position of the Department of Defense or the U.S. Government.

* This work was done while Greg Lauer was with BAE Systems.

Network Coding. DOI: 10.1016/B978-0-12-380918-6.00006-8
Copyright © 2012 Elsevier Inc. All rights reserved.

Abstract

While network coding offers many potential benefits for improving performance, practical experiences have been largely limited to simulation, emulation, and small-scale demonstrations. In this chapter we present CONCERTO, a fully implemented communication system based on a network coding protocol stack, and review observations and results from recently conducted 802.11-based MANET field trials contrasting the CONCERTO system against a candidate baseline protocol suite. These field trials were based on operational scenarios with mobile radios (pedestrian, vehicle, and airborne relay nodes) conducting a search and rescue mission using video, file transfer, chat, and situational awareness applications. The CONCERTO system was shown to support 2 to 3 times more video throughput than a state-of-the-art set of protocols, as well as up to 7 times distance-utility product. This chapter examines practical implementation issues of network coding on an embedded system and analyzes performance results.

Keywords: Network coding, mobile *ad hoc* networks, wireless networks, routing, reliable multicast, network protocols.

1. INTRODUCTION

Mobile *Ad Hoc* Networks (MANETS) have emerged as a critical component of military and first responder communication in field operations. However, despite the rise of pervasive commercial wireless communications and advances in radio technologies, MANETS continue to provide only a fraction of their potential capacity. Reasons for this include scarcity of spectrum, lack of stable infrastructure, and non-commercial requirements such as mission-critical systems that cannot tolerate "busy" signals, and the use of the layered IP-based technology that was designed for fundamentally different "wired" assumptions.

The objective of the Control-Based Mobile *Ad Hoc* Networking (CBMANET) DARPA program was to research, design, develop, and evaluate a revolutionary Mobile *Ad Hoc* NETwork (MANET) prototype that would improve effective performance from network stakeholder (user and operator) perspectives relative to the state of the art.

The BAE Systems team developed a MANET solution called Control Over Network Coding for Enhanced Radio Transport Optimization (CONCERTO) based on network coding. Network coding provides the overarching approach which: allows the network operator to achieve the full theoretical throughput capacity of wireless multicast; enables efficient solutions to multicast optimization problems that are otherwise NP-complete; provides robustness to loss and routing loops; unifies unicast, multicast, broadcast, and multi-path algorithms; subsumes rateless coding at the edges; and exploits opportunistic transmissions. Through a set of field trials, we prove the concept of a MANET communication system that is reliable, robust, and shows dramatic performance improvements under realistic mobile radio scenarios.

1.1. Challenges in Wireless MANETs

Design of reliable and efficient communication schemes over wireless *ad hoc* networks poses several challenges, especially if the IP-based set of protocols developed for wireline networks is used. First, the MANET topology can be highly dynamic, with connectivity changing in seconds or less. Any IP route based on shortest path method can become invalid and require recomputation, resulting in transmission interruptions and increased routing control overhead.

Another challenge is that radio links can easily be impaired due to physical multi-path self-interference, interference with other devices, and/or jamming. Such high and variable link losses constitute challenges to existing reliable transport protocols such as TCP and Nack-Oriented Reliable Multicast (NORM) [1] that do not distinguish between link loss and congestion loss.

High link loss is also at odds with the end-to-end principle of the current transport protocols including TCP and NORM. According to this principle, the control of reliability (checking for missing data and sending replacements) is only done at the end points (source and destination). This results in inefficiency of transmission that can grow exponentially in the number of hops in the path (since the probability of successful end-to-end reception is the product of each link's success probability).

MANETs also have the challenge of scaling to tens and hundreds of nodes. While some existing MANET protocols, such as Simplified Multicast Forwarding (SMF) [2], can overcome the above challenges for small networks (10–15 nodes), they become inefficient for networks above 20 nodes. Indeed, the strategy of flooding each data stream throughout the network, while mitigating the issues of route computation and link unreliability, is highly inefficient for the general multicast case where the destinations are a small portion of the entire network. There is also a wide range of application needs, from *unicast* (single destination) to *multicast* (multiple destinations), to *broadcast* (all nodes in the network are destinations). Current state-of-the-art protocols have separate and very different approaches for each of these cases.

All current approaches are based on today's layered IP technology which is based on fundamentally "wired" assumptions. The layered principle—hiding information between network protocols (link/MAC, routing, and transport)—has been very successful in simplifying protocol design and operation in wired networks. In contrast, in wireless MANETs, network protocols can benefit from lower layer information to provide the full potential of the network performance by overcoming the uncertainty of topology, state, and packet radio reception.

1.2. The CONCERTO Approach

The CONCERTO system is a combined routing and transport protocol that enables communication between a source and a set of destinations, which supports the entire range of applications (unicast, multicast, and broadcast) in a single framework.

CONCERTO is based on two novel paradigms that address the challenges of wireless MANETs described above. The first paradigm replaces the classic forwarding of data packets with **transmission of information via network coded data**. Application data is transformed into encoded packets by taking random linear combinations of packets from the same session (same source). The resulting encoded packets are transmitted, received, and possibly recombined at intermediate nodes. Application data is decoded at destination nodes when a sufficient number of encoded packets have been received, irrespective of their identity (this is a basic result of network coding theory, more details in Section 3). As a consequence, network coding transforms data transmission from the classical method of forwarding unique packets (that need to be identified, checked for loss and retransmitted) into information propagation where only the number of coded packets received is important, not their identity. The network coding transport module, described in Section 5, replaces the transport layer protocols such as TCP and NORM and is tightly coupled with the subgraph construction module described in Section 4.

The second paradigm of the CONCERTO system is **data propagation over subgraphs**. A subgraph is the set of nodes and links used to forward information of a session from its source to its destination(s). This paradigm subsumes IP unicast (forwarding over shortest paths) and multicast (forwarding over trees) as special cases. Subgraph construction is based on solving the equations that describe the flow of network coded information between sources and destinations. Subgraph forwarding enables simple solutions to the problems of detecting and breaking routing loops, implementing reliable multicast transmission, and exploiting opportunistic receptions (receptions by nodes other than the "next-hop")—problems which are difficult to solve in the classical IP framework. Most importantly, having only a subset of all nodes transmit packets of a session makes the CONCERTO system much more efficient than flooding. The subgraph constructor module discussed in Section 4 chooses an efficient subgraph for a given multicast session. This module replaces the classic shortest-path IP routing algorithm (used in protocols like Open Shortest Path First, OSPF [3], and Open Link State Routing, OLSR [4]) or the flooding protocols (such as Simplified Multicast Forwarding, SMF [2]).

2. CONCERTO OVERVIEW

The primary components of the CONCERTO system are (Fig. 6.1):

- Application interface (AppIF): receives packets from applications running on PCs that are connected to CONCERTO via an Ethernet, classifies them into traffic types and passes them to the network coding module.
- Network Coding module (Net Coder): groups application packets into generations, performs random linear coding over packets in a generation, provides coded packets to the forwarding engine upon request, and decodes packets at destination nodes. **This module is discussed in detail in Section 3**.
- Subgraph Constructor (SG Constructor): uses information about network topology, multicast groups, source and destination nodes to select the nodes that will forward network coded packets for each application session and to determine for each of these nodes how much traffic they should forward. **This module is discussed in detail in Section 4**.
- Forwarding engine (Master Fwder, Fwdr): uses novel transport protocols for deciding which packets need to be encoded, forwarded, decoded, discarded, retransmitted, timed-out. **Details are provided in Section 5**.
- Neighbor Discovery, Topology Discovery (Topology, Topo Cache): determines node connectivity and estimates link quality; uses the efficient Optimized Link State Routing (OLSR) protocol to disseminate connectivity information throughout the network to support the subgraph construction and to distribute multicast group membership.
- Group Manager (Internet Group Management Protocol (IGMP) Module, Group Cache): keeps track of which destinations are in each multicast group and provides this information to the subgraph constructor.

- Gateway Functionality (Protocol Independent Multicast (PIM) Module): provides interoperability between IP-based PIM multicast routing [5] and CONCERTO-based networks.

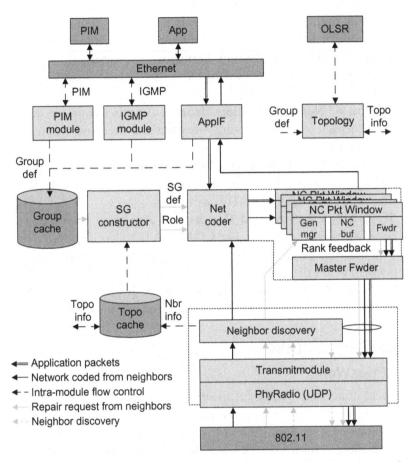

Figure 6.1 The CONCERTO architecture. Double black lines indicate the processing flow for application packets generated at this node; thick black lines indicate the processing flow for network coded packets received from neighbors (for forwarding and/or decoding); solid gray lines indicate repair request packets received from neighbors; dashed black lines indicated intra-module flow control; and dashed gray lines indicate periodic neighbor discovery packets. The light gray modules are network coding specific and are discussed in more detail in the remainder of this chapter.

The remaining sections in this chapter are organized as follows:

- **Section 3. Network Coding.** In this section we briefly introduce the basic principles of network coding and how they are applied in CONCERTO.
- **Section 4. Subgraph Construction.** In this section we provide an overview of the algorithm and module that selects the nodes and their level of participation in forwarding a session's network coded packets.
- **Section 5. Network Coding Transport Protocols.** In this section we describe the protocols used for forwarding a session's network coded packets.
- **Section 6. Network Coding Benefits.** In this section we describe the benefits of network coding—beyond the theoretical throughput improvement.
- **Section 7. Field Experiment Infrastructure.** This section describes the emulation-based approach we used to implement and evaluate both the baseline and CONCERTO algorithms.
- **Section 8. Experimental Results and Analysis.** This section discusses the performance of the CONCERTO algorithms in a set of field trials and provides insight into why the CONCERTO algorithms perform better than the baseline algorithms.
- **Section 9. Conclusions and Future Work.** In this section we discuss insights gained from research on how to further improve performance.

3. NETWORK CODING

3.1. Prior Work

The field of network coding has its origins in the work of Ahlswede *et al.* [6] and Li *et al.* [7], but its potential to transform networking has only been shown more recently. Ahlswede *et al.* [6] showed that coding within a network allows a source to multicast information at a rate approaching the smallest cut between the source and any receiver, and gave an example of a directed network where the maximum multicast throughput cannot be achieved without network coding. Li *et al.* [7] proved that linear coding with finite symbol size is sufficient for multicast connections. Koetter and Médard [8] developed an algebraic framework for linear network coding and showed that the min-cut max-flow bound of Ahlswede *et al.* [6] can be achieved with time-invariant solutions for networks with delay and cycles. They also demonstrated that network coding can be used to provide robust solutions to networks with link failures, in which only receiver nodes need to change behavior in response to failures.

Ho *et al.* [9] introduced distributed randomized network coding algorithms, in which network nodes independently and randomly choose linear mappings from inputs onto outputs over some field, and the aggregate effect of the random code choices in the network on a packet is inexpensively communicated to the receivers as a fixed-size vector of coefficients in each packet header. Reference [10] extended the randomized coding scheme to reliable communication over lossy networks. The most notable features of this scheme are: that it is capacity-achieving; that intermediate nodes perform additional coding while neither decoding nor waiting for a block of packets before sending out coded packets; that little or no feedback or coordination is required; and that all coding and decoding operations can be performed in polynomial time.

3.2. CONCERTO Network Coding

CONCERTO uses generation-based random distributed network coding (Fig. 6.2). This method was shown in [9] to achieve the capacity of a single multicast connection in a given coding subgraph (the set of network nodes participating in network coded transmission of a given data session). As a consequence, in setting up optimal single multicast sessions in a network, there is no loss of optimality in separating the problems of subgraph selection and coding, i.e., separating the optimization for a minimum-cost subgraph, which we discuss in Section 4, and the construction of a code for a given subgraph, discussed below.

Suppose that the source node s organizes a sequence of packets into *generations* of g packets, where the ith generation consists of packets $w_1^i(s), w_2^i(s) \ldots w_g^i(s)$. Now consider a node n which has received k messages in the ith generation: $w_1^i(n), w_2^i(n) \ldots w_k^i(n)$. Each message can be thought of as a vector of length ρ over the finite field $\mathbb{F}_q$. (If the packet length is b bits, then we take $\rho = \lceil b/\log_2 q \rceil$.) The coding operation performed is the same for every node: new packets are generated for transmission by taking linear combinations of the received packets where the linear coefficients are selected randomly from the field $\mathbb{F}_q$. Since all coding is linear, we can write any packet x in the network as a linear combination of $w_1^i(s), w_2^i(s) \ldots w_g^i(s)$, namely, $x = \sum_{j=1}^g \gamma_j w_j^i(s)$ where each γ_j is an element in $\mathbb{F}_q$. We call $\gamma = (\gamma_1, \gamma_2, \ldots, \gamma_g)$ the global encoding vector of

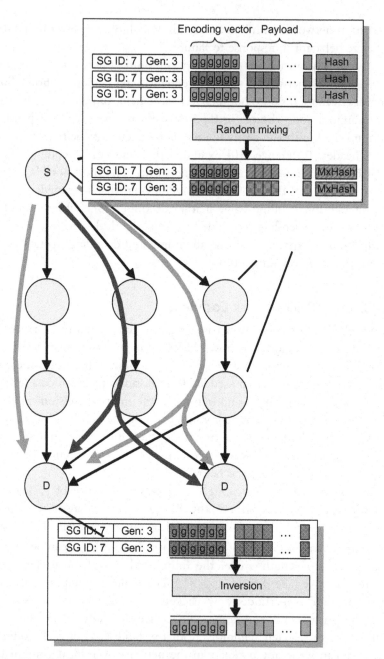

Figure 6.2 A fixed size vector in each network coded packet contains the linear coefficients associated with the source packets. These coefficients are used by the destination to decode a generation of packets.

x, and we assume that it is sent along with x, in its header. Note that intermediate nodes do not have to collect g packets before they can transmit, thus reducing network latency. The overhead this approach incurs (namely, $g \log_2 q$ bits) is negligible if packets are sufficiently large. For example, a generation of size 10 and field of size 2^8 (typical values for CONCERTO) yields an overhead of 10 bytes which is negligible compared to data payload of 100 to 1000 bytes.

A sink node collects packets and, if it has g packets with linearly-independent global encoding vectors, it is able to recover the message packets in that generation. Decoding can be done by Gaussian elimination. Our implementation of network coding can be summarized as follows:

- The source node collects packets into groups of g packets (generations).
- Each node in the subgraph transmits random combinations of packets in a generation.
- Packet headers contain an *encoding vector* consisting of the coefficients which summarize the random coding performed to date.
- Each destination collects packets until it has g linearly independent coding vectors.
- Each destination inverts the matrix of coefficients extracted from the g network coded packets to recover the original packets.

The encoding vectors of packets received by a node for a given generation define a subspace in $\mathbb{F}_q^g$, which we call the *received subspace*. A packet is *innovative* if its encoding vector does not lie in the received subspace and is *non-innovative* if it does. The *null space* is the subspace of $\mathbb{F}_q$ that is orthogonal to the received subspace.

With high probability, distributed random network coding achieves the single multicast capacity of a subgraph. The details of how we construct this subgraph are given in Section 4.

4. SUBGRAPH CONSTRUCTION

4.1. Algorithm

Only a subset of the MANET nodes participate in generating network coded packets for a given multicast session. Subgraph construction refers to determining this subset and, importantly, the rate at which these nodes transmit network coded packets. It is possible to formulate the subgraph construction problem as a linear programming (LP) problem (e.g., [11]), and an early version of the CONCERTO algorithms did just that. How-ever, the number of constraints in the LP problem is exponential in the number of multicast sinks and thus does not scale to large networks.

The CONCERTO subgraph construction algorithm is based on the MORE algorithm ([12]). The subgraph includes all nodes that are closer to any of the multicast destinations than the source, where distance is measured using the ETX metric (the expected number of transmissions to get to the destination). Each node in the subgraph generates and transmits a *forwarding factor* number of network coded packets whenever it receives an *innovative* packet. The forwarding factor is computed by the source node for each node in the subgraph and is included in the header of each packet injected into the subgraph. This allows the source to re-compute the forwarding factor whenever the topology changes or whenever link qualities change. To reduce channel access contention, nodes that have a forwarding factor smaller than a threshold are not included in the subgraph.

While the CONCERTO subgraph and forwarding factors are computed using the MORE algorithm, we used original reliable transport mecha-nisms to determine when additional transmissions are required and when to stop sending network coded packets. These mechanisms are discussed in Section 5.

4.2. Implementation

Figure 6.3 illustrates how the Subgraph Constructor interacts with the other components in the CONCERTO system.

Group Manager

Each multicast session uses a unique multicast IP address. The Group Man-ager uses the IGMP protocol [13] to gather information about which nodes are in each multicast group. It uses the OLSR protocol [4] to flood the following information:

- group ID;
- source node;
- destination nodes;
- application type: elastic fully reliable (file transfer), inelastic semi-reliable; (streaming media, chat, Situation Awareness—SA).[1]

---

[1] For simplicity, different IP address ranges are used for each type of service.

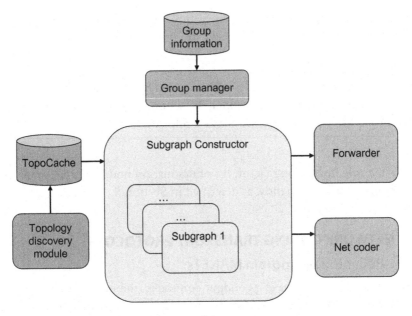

Figure 6.3 Subgraph constructor system diagram.

Topology Discovery Module

The Topology Discovery Module dynamically detects nodes and determines link quality using beacons and overheard traffic from neighbor nodes. It stores this information in the Topology Cache. Remote link quality information is provided by the link state dissemination protocol of OLSR. No other function (such as shortest path routing) is used from OLSR.

Topology Cache (TopoCache)

An object database that contains all the nodes in the network and the links that connect them along with their associated link quality information.

Subgraph Constructor

The Subgraph Constructor iterates over Group Records and calculates subgraph information. It uses information from the TopoCache (which contains information about the nodes in the network, the links, and the probability of loss associated with each link) and from the Group Manager (which maps multicast addresses into type-of-service requirements). The Subgraph Constructor produces information which is consumed by the NetCoder and Forwarders which implement transport level reliability mechanisms.

Net Coder

Uses the subgraph ID and traffic type provided by the subgraph constructor to determine that a Forwarder should be set up for a given subgraph (more precisely a Slave Forwarder, see Section 5 for details).

Forwarder

Uses the role, forwarding factor, list of destination nodes, and subgraph ID to transmit traffic efficiently, as described in Section 5.

5. NETWORK CODING TRANSPORT PROTOCOLS

5.1. Reliable Transport in MANETs

Applications typically have a goodput constraint (the minimum amount of information that must be received) and a latency constraint (a time after which the information is no longer useful). Existing reliability methods developed for wireline networks exhibit poor performance in MANETs due to the network's highly dynamic nature (mobile nodes and links with high and variable losses). We have developed two reliable transport protocols that are appropriate for use in unreliable MANETs. These protocols leverage the features of network coding to provide different trade-offs between goodput and latency.

- **Semi-reliable.** Applications such as interactive voice or streaming video can tolerate a small percentage of missing packets. However, they require that packets be delivered within a defined latency bound to be useful. Moreover, these applications can recover after a period of high loss. Our semi-reliable algorithm uses a hop-by-hop mechanism and aggressive retransmit policies to ensure that a high fraction of packets are delivered to all receivers within a configurable latency bound (typically 100s of milliseconds to a few seconds).
- **Fully-reliable.** Applications such as file transfer or email require that all packets constituting a data element be delivered—any missing data renders the whole transfer unusable. Latency requirements for such applications are typically more relaxed than for interactive applications. Our full reliability algorithm uses end-to-end mechanisms in addition to

hop-by-hop mechanisms in order to reduce the probability that a complete data element is not delivered to a destination—despite network events such as network partitioning, severe packet loss, etc.

In the following we describe the protocol architecture that defines the feedback loops used to generate and transmit network coded packets in a subgraph, the master/slave architecture that determines which subgraph's packet is scheduled next, and the slave forwarder algorithms for the Semi-Reliable and Fully-Reliable Transport Protocols. Parts of these algorithms were inspired from recent work on CodeCast [14].

5.2. Forwarding Protocol Architecture

CONCERTO establishes a network coded multicast session in which the source node is connected to the receiver nodes via a directed subgraph of nodes. In this directed subgraph, each node has downstream neighbors (except for destinations that are not forwarders) and upstream neighbors (except for the source). At the source, packets from application are first grouped in generations and then random linear combinations of packets in each generation are broadcast to its downstream neighbors. Each downstream node receives network coded packets, then combines those belonging to the same generation, and broadcasts new random linear combinations to its downstream nodes in the subgraph. The destinations decode the original packets when they have received enough linearly independent encoded packets from a generation.

To provide efficient and time-effective loss recovery, our algorithm includes local repair strategies. Unlike end-to-end algorithms such as TCP, local repair uses packets temporarily stored at intermediate nodes (necessary for network coding) to generate new network coded packets, without the need for end-to-end feedback. A node can receive local repair packets from any of its upstream neighbors, which increases the likelihood of repairing local losses.

In the following we illustrate the protocol using an example network (Fig. 6.4) in which there is a multicast session from a sender node S to destination nodes D and E. The session's subgraph includes forwarding nodes A, B, and C. Packets received from an application at the sender node are grouped into generations. Network coded packets are then generated and transmitted over the wireless channel using local broadcast (solid arrows). To

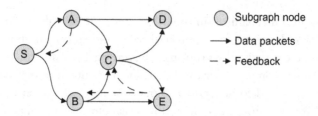

Figure 6.4 A six-node network that shows the flow of data and reliability control information.

first order, the number of coded packets sent by a node is determined by the *forwarding factor* computed by the Subgraph Constructor. However, the forwarding factor compensates for the *average* number of losses on the wireless channel, not the actual number of losses experienced per generation.

If a node receives too few coded packets in a generation within a given time, it generates a feedback request indicating that more innovative packets from that generation are needed (dashed arrows). To avoid having the upstream node send non-innovative packets, the feedback request includes information about the requesting node's null space (packets in the null space are innovative for that node at that time). Whenever possible, the feedback is *piggybacked* onto the header of any data packet ready for transmission since broadcasts for downstream neighbors can typically be overheard by upstream neighbors. It would be expensive to include the whole null space in every packet transmission, so only a random vector from the null space is included in the feedback request.

Upstream nodes that receive a feedback request check to see if the random null space vector is in their receive spaces. If it is, they respond to that request by transmitting a random linear combination of packets from that generation. If not, they ignore the feedback request since with high probability they cannot generate an innovative packet. To improve efficiency, the transmission intensity with which a node responds to a feedback is equal to its local forwarding factor, described above. Transmit requests can be redundantly included on multiple consecutive packets, thus enabling feedback delivery with high probability even at high levels of packet loss.

Finally, we observe that this algorithm decides only on which packets and feedback information to transmit, leaving the timing of their transmission to be decided by a rate control (QoS) module.

5.3. The Master/Slave Architecture of the Net Coding Transport Protocols

A node can contain packets from different subgraphs and from different generations within a subgraph. The *master/slave* architecture (Fig. 6.5) determines which subgraph should send next and which generation within a subgraph should be sent. It also determines whether repair requests need to be piggybacked onto these transmissions or whether an explicit repair request must be sent. Specifically, each node has a slave forwarder for each subgraph that receives data and control information (feedback) for that subgraph, and that chooses the generation (data and/or control) to send next. A *master forwarder* arbitrates between all sessions existing on a node and chooses

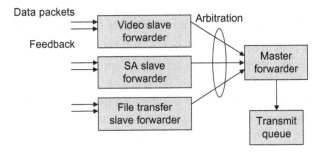

Figure 6.5 The Master/Slave forwarding architecture.

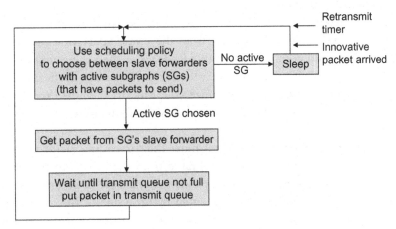

Figure 6.6 Algorithm used by the Master Forwarder to select the next packet to send.

the next slave forwarder to send. The master forwarder uses a priority-based policy, where sessions with weaker delay constraints such as file transfer are selected for transmission only when there is no packet available to send from the tighter delay sessions such as voice and video.

The Master Forwarder chooses the highest priority session that has a packet to send and queries its associated slave forwarder for a packet and feedback information from that session (Fig. 6.6). It tries to put that packet in the transmit module's input queue and, if that queue is not available, waits (blocks) until a packet is transmitted and space is available. If there is no packet ready to be sent from any session, then it sleeps until it is reactivated either by an innovative packet received by any session, or by a timer indicating that a session retransmission is required.

5.4. The Semi-Reliable Slave Forwarder Algorithm

The semi-reliable transport algorithm is designed to:

- Minimize the number of packet transmissions per unit of application data.
- Minimize the time to transfer a full generation from source to destinations. This minimizes application end-to-end delay and minimizes the number of generations in the network.
- Minimize the delay between packet construction and packet transmission, so that the information contained in a transmitted packet is current and useful to downstream nodes.

The semi-reliable slave forwarder assembles the next packet to send by generating a network coded packet from the oldest generation that has a packet to send and combining it with a feedback request indicating the set of all generations that need to receive more packets from upstream. The oldest-generation-first policy minimizes the time to complete the transmission of a generation and thus minimizes the end-to-end transmission delay.

The subgraph computation determines the forwarding factor used by each node in the subgraph to move traffic from the source to the destinations. The forwarding factor is implemented via a credit counter mechanism. The credit counter is incremented by the value of the forwarding factor upon each innovative packet reception and is decremented by one for every packet sent, but limited to not become negative. If the credit is smaller than one, a packet is made available with probability equal to the credit. In the case of the source node, a complete generation is first assembled from the application and then the credit counter is initialized with a credit equal to generation size times the forwarding factor.

Besides the transmission of packets prompted by upstream innovative data arrival, there may be a need for more transmissions as more packets can be lost on the wireless channel than the redundant set provided for by the forwarding factor. Nodes that determine they have insufficient information (i.e., they have a generation that is not full rank) request local repair packets via a feedback message (that is piggybacked onto the data transmission when possible). Such a repair request is generated if no innovative packet is received within a configurable time. While feedback-based transmission is relatively efficient (packets are only sent if they are needed), it does introduce delays. Our algorithm allows us to adjust the parameters to trade-off between delay and efficiency.

As repair requests can potentially be received from many nodes within a short time, there is a possibility of sending redundant repair packets, since one packet can satisfy multiple requests (this is the well-known FEC-like property of network coding). To maximize efficiency, this algorithm uses a timer to accumulate such requests and produce one repair packet.

5.5. The Fully-Reliable Slave Forwarder Algorithm

Some applications, such as file transfer or email, require that all the packets in a file (or other data unit) be received by a destination for them to be of any value. Many of these applications have relatively loose delay constraints and thus can benefit from the use of an end-to-end reliability protocol.

The semi-reliable forwarding algorithm described above provides recovery from many cases of lost packets based on detection of generations that do not have full rank. However, that algorithm does not attempt to determine if a node had missed generations before it joined the subgraph (since resending old generations is not useful for latency sensitive applications) and does not implement mechanisms to handle certain special cases (e.g., no packets received in the last generation).

The fully reliable protocol extends the hop-by-hop semi-reliable protocol with methods to detect and recover missing generations. The source node uses heuristics to detect the start and end of file transfers and injects markers representing these events into the network coded packet stream. Forwarder and destination nodes send repair requests when they detect missing generations. These repair requests are propagated upstream until some node (possibly the source) is reached that can provide packets in that generation.

If a destination node joins a multicast session midway through a file transfer, the source node might have to retransmit a significant amount of data—perhaps more than can be sent in time to meet even the lax file transfer latency requirement. To handle this case, the fully-reliable forwarder can be configured to limit how long it saves generations.

6. NETWORK CODING BENEFITS

A powerful feature of network coding is that it subsumes a wide range of other MANET mechanisms. This means that network coding provides an infrastructure that unifies hitherto disparate network algorithms—greatly simplifying the configuration and coordination of these algorithms.

6.1. Unified Broadcast, Multicast, and Unicast

Network coding subgraphs are constructed between a source and a *set* of destinations. Since the destination set can be a single destination, all the other nodes in the network, or a subset of the nodes in the network, network coding provides a single mechanism for implementing unicast, broadcast, and multicast. Figure 6.7 illustrates a subgraph that supports a flow of information from the source node S to destinations D_1 and D_2.

6.2. Robustness to Routing Loops

In MANETs, the latency in disseminating topology data means that nodes may not have an accurate picture of the network topology. In packet forwarding networks this can lead to a problem with forwarding loops. For example, consider the scenario depicted in Fig. 6.8 where a forwarding loop exists. In networks using (traditional) routing these loops cause significant problems—possibly making destinations unreachable and reducing bandwidth through useless retransmissions of the same packet. Since network coding forwards information flows on a subgraph

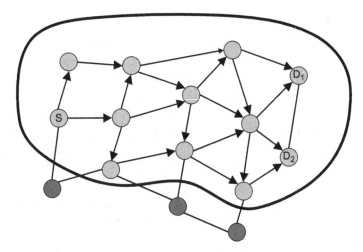

Figure 6.7 The network coding problem formulation treats unicast and broadcast as special cases of multicast. Here the light gray nodes are part of the subgraph that forwards information from the source S to the destinations D_1 and D_2.

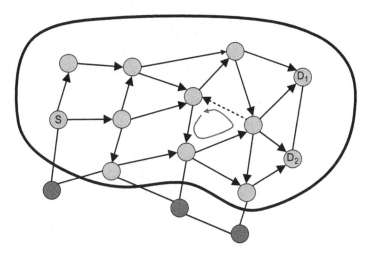

Figure 6.8 Network coding is robust to routing "loops" since packets that are not innovative (do not contain new information) do not trigger a transmission.

(via mixtures of packets in a generation) rather than packets, network coding does not have a "forwarding loop" problem: if a packet is received which is not innovative then it does not lead to any further transmissions.

6.3. Robust to Link and Node Failures

Note that subgraphs are typically robust to link or node failures since the optimal solution frequently incorporates more nodes than would be used in a routed network's spanning tree. In routed networks the use of single paths to forward traffic causes fragility (e.g., the loss of a node or link can cause a destination to become unreachable). This fragility can be reduced if the source computes and uses multiple link-disjoint or node-disjoint paths between itself and the destination. However, this approach is complex and does not extend easily to multicast traffic. Since network coding forwards information along all the paths in the subgraph, a k-connected subgraph will be resilient to failures of up to $k - 1$ links/nodes—and does so for both unicast and multicast traffic without the complexity of determining link/node disjoint paths.

6.4. Provides Low-Latency Link Layer Coding

Forward Error Correction can be used to improve reliability. Packets are encoded so that, for every group of K packets to be transferred, N coded packets are sent. The code is constructed so that if K or more encoded packets are received, then the original K packets can be reconstructed, regardless of which packets are received. The values of K and N are selected based on packet loss probability, reliability requirements and latency constraints. If source-to-destination coding is used (e.g., NORM [1] or Digital Fountain [15]) then the capacity is simply the product of the probabilities that a packet is successfully delivered over the links along the path. In Fig. 6.9, which illustrates a two-hop path, the capacity would be: $(1 - \varepsilon_1)(1 - \varepsilon_2)$ where ε_i is the probability of packet loss on link i (assuming for simplicity that all links have the same capacity). If per-link FEC coding is used to reconstruct packets at each intermediate node, then the

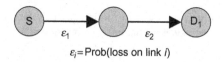

ε_i = Prob(loss on link i)

Figure 6.9 Network coding provides low-latency link layer encoding as part of its normal operation.

capacity is higher: $\min(1 - \varepsilon_1, 1 - \varepsilon_2)$, but delay is increased since each packet must be reconstructed before it can be forwarded.

Network coding achieves the optimal cut-set capacity (in the example of Fig. 6.9, this is $\min(1 - \varepsilon_1, 1 - \varepsilon_2)$), however packets are not reconstructed at intermediate nodes, and thus network coding attains the lower latency of end-to-end encoding where packets only need to be decoded at the destination.

In MANETs the loss characteristics of links may change too rapidly to be characterized accurately. In this case a rateless approach to link layer coding may be required in which a sequence of encoded packets is generated until the destination has recovered the source packets. End-to-end approaches such as Digital Fountain apply this approach for multicast networks. Network coding provides this functionality at a hop-by-hop level—each node can send random linear combinations until downstream nodes have enough information—with the expected benefits of reduced transmissions and increased capacity. Feedback is required to let upstream node(s) know when to stop sending and the rate at which they should send. Algorithms for accomplishing this were discussed in detail in Section 5.

6.5. Extremely Opportunistic Routing (ExOR)

Biswas and Morris [16] observed that we generally do not know in advance which nodes will receive a radio transmission. Thus the typical routing approach in which the destination for a packet is pre-specified is inefficient—it would be better to transmit the packet, determine which nodes have received it and then decide which nodes should forward it. Based on this observation they developed the Extremely Opportunistic Routing protocol (ExOR). This protocol dynamically determines the sequence of nodes to be used to forward each packet—based on information about which nodes actually received each packet. The node closest to destination (using the Expected Transmit Count (ETX) metric) is selected to transmit the packet—moving the packet farther (on average) than is possible with the best possible predetermined route. This algorithm focuses on unicast traffic and is not easily extended to support multicast traffic.

Two motivating examples were given in [16] to illustrate the gains available from ExOR. In the following, we discuss how network coding subsumes opportunistic routing—and extends it to multicast traffic.

6.5.1 Long Hops

Frequently the probability of a node receiving a transmission is a function of the distance from the transmitter. In these cases, transmissions may have a high probability of being received by close-by neighbors and a lower probability of reception by more distant neighbors. Typical routing protocols require that links only be used if their link quality is sufficiently high (since the number of retransmissions at the link layer is otherwise excessive). This means that links that infrequently move a packet a long distance may not be used—reducing the efficiency of the network. Since ExOR determines transmission order after nodes have received packets, it can exploit these long low-probability links to its advantage (Fig. 6.10).

Network coding also supports the use of long hops in a natural fashion. If all the nodes in Fig. 6.10 are part of the subgraph, the rate at which each node participates will be determined by the quality of the links between nodes in the subgraph. When a distant node opportunistically receives a network coded packet it simply adds it to the pool of information it has about that generation—which decreases the time before the destination can decode. Note that network coding provides the "long hop" advantage of ExOR to multicast traffic as well as to unicast traffic.

6.5.2 Lots of Lossy Links

In traditional routing, a single "next-hop" is selected out of all the possible next hops to a destination. The choice of next-hop is typically selected to minimize the "distance" to the destination. If there are many alternatives—all of which are poor—this strategy can be substantially suboptimal. For example, Fig. 6.11 illustrates a scenario in which there are 5 alternative next-hops between a source S and a destination D_1. The probability of successful reception by each of these nodes is 20% so, even though any

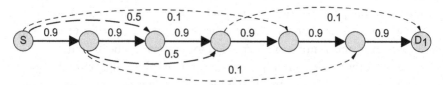

Figure 6.10 Network coding naturally incorporates opportunistic receptions since innovative packets are always stored upon reception.

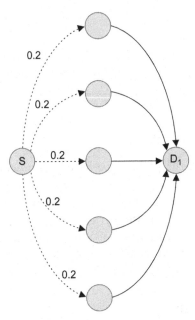

Figure 6.11 Network coding subgraph construction takes into account link quality and uses multiple low quality links if that is optimal.

given node is quite unlikely to receive a transmission, it is very likely that at least one of these potential next-hops will receive it.

ExOR takes advantage of multiple lossy links by determining which node transmits *after* the packet is received. ExOR would use all the intermediate nodes in this figure for forwarding. Network coding also exploits multiple lossy links in the same manner—but does so in a way that supports multicast traffic as well.

In summary, network coding subsumes a wide range of algorithms that hitherto have been considered as solutions to distinct problems: unicast, multiple-path and multicast routing, link layer coding, opportunistic forwarding.

7. FIELD EXPERIMENT INFRASTRUCTURE

The goal of our experiments was to evaluate the *real-world performance* of the CONCERTO protocol suite running on 802.11 radios (and compare it against a baseline set of protocols) through a series of field experiments in a MANET environment designed to support mobile nodes using a portable embedded platform.

In this section we describe the hardware and software applications and tools used in our field experiments, and the corresponding criteria used to evaluate on-field performance. We also present a brief overview of the baseline suite of protocols.

7.1. Hardware

During the course of the project, we have experimented with three platforms: "Avilla"—using a single-core 533 MHz ARM processor; "Aeon"—using a dual core 2 GHz Intel; and "Pismo"—using dual core 1 GHz ARM. In all trials of record we used "Pismo" given its good balance between CPU power and low power consumption (8 W), with no special cooling. We used commercial off-the-shelf (COTS) 802.11a/b/g wireless radios using the Atheros chipset with characteristics shown in Table 6.1.

7.2. Baseline System

Our evaluation methodology consisted in comparing our CONCERTO system with a state-of-the-art baseline system under similar network and traffic conditions. The baseline system (Fig. 6.12) consists of an 802.11 wireless radio running Optimized Link-State Routing (OLSR) [4], Basic Multicast Forwarding (BMF) [17] (similar to Simplified Multicast Forwarding [2]), and NACK-Oriented Reliable Multicast (NORM) [1]. ALARES (Application Layer Reliability System [18]) was used to provide recovery of data lost during longer-term network outages. These protocols were selected and tuned with the help of a government-selected neutral test team.

OLSR is an IP routing protocol designed to be efficient in MANETs. Each node generates and floods topology messages that list its neighbors and

Table 6.1 Characteristics of the 802.11 radio

Radio hardware	Ubiquity 802.11 wireless card (XR-2)
Tx power	600 milliWatts (28 dBm)
Transmit data rate	11 Mbits/second
Receive sensitivity	−92 dBm
Ground node's antenna	12″ omni (+7 dBi)
Airborne node's antenna	12″ omni (+7 dBi) plus 120 degree sector (+9.5 dBi)

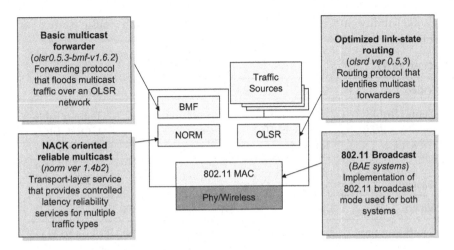

Figure 6.12 Architecture of the baseline system.

the link quality to each neighbor. Unlike other routing protocols, flooding uses only a subset of the nodes (known as multipoint relays), which increases efficiency of using network capacity.

The NORM protocol is designed to provide end-to-end reliable transport of bulk data objects or streams over generic IP multicast routing and forwarding services. NORM uses a selective, negative acknowledgement (NACK) mechanism for transport reliability. The protocol leverages the use of FEC-based repair to allow a single retransmission to help multiple destinations regardless of which packets they have missing. FEC can optionally be enabled to provide proactive transmission robustness. Video latency requirements (4 seconds) were not compatible with the use of NACKs and thus, for video, NORM was configured with proactive rate ½ FEC coding. For file transfers, NORM was configured with a proactive rate ½ FEC code and NACKs.

ALARES is designed to provide longer-term reliability in MANETs in a manner that is complementary to FEC and NORM. ALARES mainly addresses temporary network partitioning where the source of information is not reachable by the destinations (e.g., when NORM would not work). ALARES stores overheard transmissions at each node and elects repair servers to provide information to destinations when the source is temporarily disconnected.

Reliability mechanisms were configured for each traffic source as follows:

CONCERTO

Video	Chat	SA	FX
Semi-reliable CONCERTO forwarder	Semi-reliable CONCERTO forwarder	Updates sent once a second	Fully-reliable CONCERTO forwarder

Baseline

Video	Chat	SA	FX
NORM set to ½ rate code, no NACKs	Each packet sent 10 times and ALARES	Updates sent once a second	NORM set to ½ rate code with NACKs and ALARES

7.3. Scenario Traffic

All field experiments were performed under similar conditions using both CONCERTO and baseline systems. We used a scripting tool ("Nettion" [19]) to control the field experiments and to ensure that the same traffic patterns were used for both CONCERTO and baseline. CenGen Inc. provided the software applications for Video, Chat, Situational Awareness (SA), and file transfer (FX), which replayed or generated real application traffic or recorded video streams. Meta-data was piggybacked on application traffic to enable measurement of various metrics such as packet losses and latency during post-processing. The data rates associated with these traffic sources are summarized below.

Application Traffic

Video	Chat	SA	FX
100Kb/s MPEG4, 12pkt/s, 1024B/pkt	2–400B/pkt Average 1 pkt/30sec	200B/pkt Average 1 pkt/sec	Average file size of 1MB

7.4. Evaluation Methodology

Our goal was to test and evaluate CONCERTO in the field and compare its performance with the baseline system based on the utility to the end users. Utility is defined on a per application basis to account for the impact of delay and packet loss. The video metric captures the notion that a video receiver is satisfied with a 10 second chunk of video if it receives 90% of the packets in that chunk with less than 4 seconds of end-to-end delay. We also observe that, while video streams have higher value if delivered over longer distances, in a mobile multi-hop network they are more difficult to transport. To provide a better sense of "transport utility" in a MANET, we designed a Distance-Utility metric, defined as the sum of the per source-destination video utility multiplied by the distance between each source and each destination.

The file transfer utility metric captures the notion that a file transfer receiver is satisfied if it receives the complete file within a certain delay limit. The file transfer utility metric is the fraction of receivers that are satisfied averaged over the different file transfers. Since file sizes did not vary significantly, the file transfer delay limit of 15 minutes was used for all files.

Each Chat message is sent to all destinations. The "value" of receiving a packet is 100% if it is received within 2 seconds; otherwise the value decays exponentially with the time between transmission and reception. The exponential decay time constant is 3 hours. The scenario Chat utility is the total value of the received packets divided by the number of possible receptions.

Each Situation Awareness (SA) message is sent to all destinations. The SA application is satisfied if it receives an update every 5 seconds. Its utility value score decays exponentially with a time constant of 100 seconds if it does not receive an update for more than 5 seconds. The scenario SA utility is the average of all time slices under consideration for that phase.

8. EXPERIMENTAL RESULTS AND ANALYSIS

In this section, we present results and analysis of the performance benefits of CONCERTO over the baseline protocol suite in dynamic and mobile tactical scenarios. We also present results from several engineering variants of the tactical scenarios that will help the reader understand the potential strengths of our system in other, more general environments.

8.1. Experiment Scenarios

The tactical scenarios were designed to showcase CONCERTO performance in a Wireless Mobile *Ad Hoc* Network and involved 35 wireless nodes (Fig. 6.13). These nodes included three teams (Alpha, Bravo, and Charlie) using 31 battery-powered nodes that were carried by individuals on portable backpacks, and four other wireless nodes mounted on two trucks and two planes. The two trucks were used to deploy and collect personnel to and from the central command post.

The scenario simulated an attack on 3 compounds by the three teams. The scenario begins with the forces assembled at a Landing Zone (LZ) near the Command Post. A small team patrols the command post. Squad members deploy static nodes in key locations, and then are deployed to surround their objective sites. The respective teams then proceed to "attack" their target sites by walking around them. The attack phase was broken into two parts: the Alpha/Bravo phase where those teams walked around their target while the Charlie radios were stationary; and the Bravo/Charlie phase where the Alpha radios were stationary. The vehicles traveled up and down on access roads carrying personnel and supplies. The different phases of the tactical scenario are described in Table 6.2; the Alpha/Bravo portion of the scenario is illustrated in Fig. 6.13.

Two scenarios were executed: a ground scenario with no aircraft, and an air scenario in which two airplanes each equipped with a radio and two high-gain antennas circled overhead acting as relays to provide shorter paths between ground units.

The scenarios included many poor quality links. Figure 6.14 illustrates the topology of the ground scenario during the Bravo/Charlie phase. Links are coded with two colors to indicate their quality as measured by each end of the link. Note the large number of links with packet loss exceeding 30% and the presence of numerous asymmetric links.

Table 6.2 Phases of the scenario used to evaluate CONCERTO and baseline performance

Phase	Action	Time
LZ	Collect at Landing Zone (LZ)	0–30 minutes
Deploy	Alpha, Bravo, and Charlie Teams Deploy	30–40 minutes
Alpha/Bravo	Alpha and Bravo Teams Attack	40–70 minutes
	Switch Personnel	70–110 minutes
Bravo/Charlie	Bravo and Charlie Teams Attack	110–140 minutes
Engineering	Stress testing	140–150 minutes

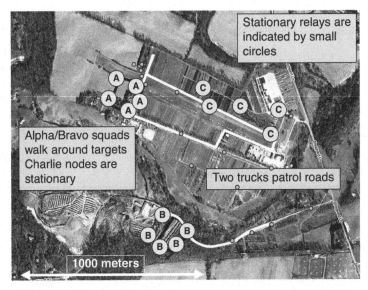

Figure 6.13 Alpha/Bravo Phase includes 10 man-pack radios, two truck mounted radios, and two aircraft mounted radios (not shown) as well as 21 stationary radios.

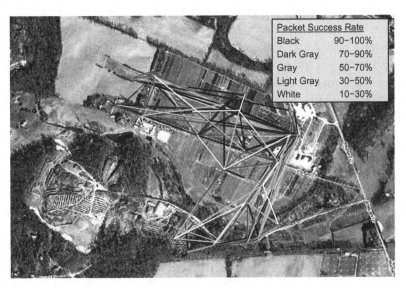

Figure 6.14 MANET topology during the Bravo/Charlie phase in the ground scenario. Note the large number of poor quality links.

Table 6.3 Video loads used to evaluate performance

Video Load	Comments	Total Receive Traffic
Load 1	3 Intra-team video sessions	1200 kilobits/second
Load 2	Video Load 1 *plus* one inter-team video session	1600 kilobits/second
Load 3	Video Load 2 *plus* a second inter-team video session	2000 kilobits/second
Load 4 (Stress Test)	Video Load 3 *plus* a third multi-hop video session	2400 kilobits/second
Load 5 (Stress Test)	Video Load 4 *plus* a fourth multi-hop video session	2800 kilobits/second

Video traffic was generated by having source nodes multicast 100 Kilo-bit/second MPEG4 streams to different sets of four destinations. In order to find the maximum loads that the CONCERTO and baseline systems could handle, three different video loads were applied during the LZ, Alpha/Bravo, and Bravo/Charlie phases as shown in Table 6.3. The three video sessions in Video Load 1 were each from a source in the team to the remaining four members in that team. Since the members of a team are typically within one-hop of each other, this is a relatively unstressful scenario. In Video Loads 2 and 3, the additional sources send to two destinations in the other clusters and to two other distant locations—a much more stressful multi-hop scenario. The third column indicates the total amount of video traffic that could be received summed over all the destinations. Since each video session has four destinations, the total amount of received traffic is just four times the number of video sessions times the video bandwidth (100 kilobits/second). This is the amount of traffic that could be received if the MANET is working perfectly.

Additional video streams were added during an Engineering Stress Test designed to determine the limits of the CONCERTO protocols. Each additional video stream was multi-hop and had four destinations.

Figure 6.15 illustrates the video flows. The sources for Video Load 1 are the circled nodes within the clusters near the three objectives. The additional session in Video Load 2 is within the black circle and the black dashed lines indicate its destinations. Video Load 3 is indicated by white

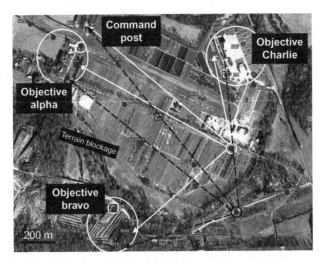

Figure 6.15 Video flows used to evaluate baseline and CONCERTO performance in ground and air scenarios. Video Load 1 is "intra-squad" (within the white circles) while Video Loads 2 (black dashed lines) and 3 (white lines) add "inter-squad" multi-hop sessions.

lines. Note that the additional sessions in Video Loads 2 and 3 involve destinations that are much further away than for Video Load 1.

File transfers (FX) were executed between 7 nodes (one FX source in each cluster, two trucks, the command post, and a sensor node). Files were transferred sequentially from each of the nodes to all remaining FX Nodes. Chat and SA traffic was generated and received by all nodes.

8.2. Experimental Results

All experiments contained a mix of video, Chat, SA, and file transfer. Chat and SA represented a relative small portion (<10%) of total traffic. Since they have robust application–level reliability algorithms, their average utility was high (above 90%) under both baseline and CONCERTO during all phases of experiments. Below we discuss results for video and file transfer.

8.2.1 Ground Tactical Scenario

Figure 6.16 summarizes the video performance of the baseline protocols (top) and CONCERTO protocols (bottom) in the Ground scenario. The

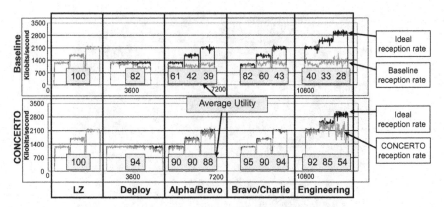

Figure 6.16 Video performance comparison CONCERTO vs. Baseline in the Ground scenario.

black lines plot the total amount of video traffic that is possible to be received with perfect performance (see Table 6.3). The gray line plots the *actual* total video traffic received by the destinations. Recall that video utility is defined as 1.0 if a destination receives 90% of the packets in "chunk" with latency less than 4 seconds, and 0.0 otherwise. The average video utility for each video load is given in the pale gray boxes.

At the start of the experiment (Landing Zone phase) nodes are stationary and are in a mesh configuration. Both CONCERTO and baseline had nearly perfect video utility metrics (average 99%) for all three video loads. In the next phase, where squad members and fixed nodes are deployed, CONCERTO performed slightly better than baseline (average utility of 94% vs. 82%). The baseline exhibited minor problems as the nodes got further apart toward the end of the deployment even though the video load was low (Load 1).

In the rest of the phases, CONCERTO performed well: the average utility for CONCERTO in each test ranged between 88% and 95%, while the baseline was effectively unusable (averaging between 30% to 60%). The main reason for CONCERTO's good performance was that the network coding subgraphs are resilient to mobility and its reliable end-to-end transport mechanisms used network bandwidth efficiently, allowing it to support more video sessions.

CONCERTO's video latency (Fig. 6.17) was comparable to that of the baseline's in all of the scenarios and was typically under 1 second. While

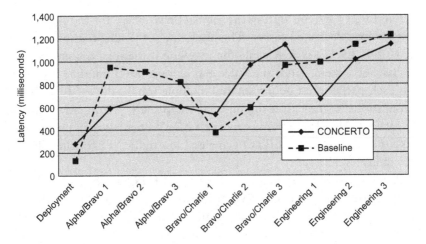

Figure 6.17 CONCERTO and baseline latency for the air scenario.

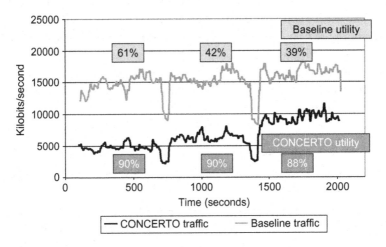

Figure 6.18 Comparison of total network transmission.

CONCERTO latency was higher than baseline latency in some phases, this is due to the fact that the baseline algorithms were dropping most packets and latency was only measured on the packets that were received by destinations. The higher CONCERTO latency is associated with performance that is much better than the baseline.

Moreover, this high performance was produced with lower network load. Figure 6.18 illustrates the performance of the baseline and

CONCERTO systems during the Alpha/Bravo phase for the three levels of video load described earlier. The black line is the total transmissions for the CONCERTO system while the gray line is the total transmissions by the baseline system. Since the video traffic must travel multiple hops (and thus each packet must be transmitted multiple times), the total number of transmissions is larger than the injected video traffic. Note, however, that the total CONCERTO traffic generated is 2–3 times less than the baseline. Note that while CONCERTO generated less traffic, it performed better: 90% utility vs. 47%.

A very important benefit of CONCERTO is that high video utility is delivered irrespective of distance. Figure 6.19 plots the average video utility perceived by a destination node as a function of its distance from its source. We observe a stark difference between CONCERTO, which delivers satisfactory video at all distances, and the baseline, which is only able to perform satisfactorily when destinations are one- or two-hops away. The main reason for this performance is the combination of network coding with the hop-by-hop reliability that is built into the network coding forwarding algorithm.

To quantify the benefit of delivering video over distance we introduce the *distance-utility metric*, which we define as the sum over all destinations of each destination's utility score multiplied by the physical distance from

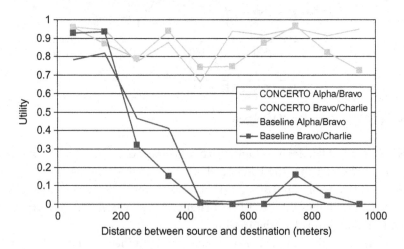

Figure 6.19 Performance over distance.

the source to the destination. The maximum distance-utility metric is computed assuming all destinations receive traffic with perfect utility. Figure 6.20 plots (for the ground scenario) the maximum distance-utility metric and the actual values achieved by the baseline and CONCERTO protocols. In our tactical experiments, Video Load 1 only went "one-hop" to receivers in their own squads, while Video Loads 2 and 3 traversed multiple hops, leading to large increases in video distance-utility for those loads. Observe that CONCERTO provides up to a 7 times gain in distance-utility metric over the baseline, and frequently achieves the maximum distance-utility metric. During the Engineering stress tests, the baseline protocols collapse while the CONCERTO protocols continue to work reasonably well, even when confronted with more traffic than can be handled.

Part of the reason that CONCERTO performs so well is that its subgraphs provide a rich set of alternative paths to the destinations while efficiently modulating the rate at which nodes forward traffic. Figure 6.21 is a snapshot of a ground scenario subgraph: the nodes are colored according to their role: black is the source of the multicast session, gray

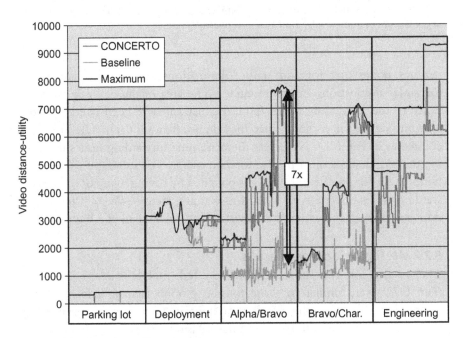

Figure 6.20 Distance-utility comparison.

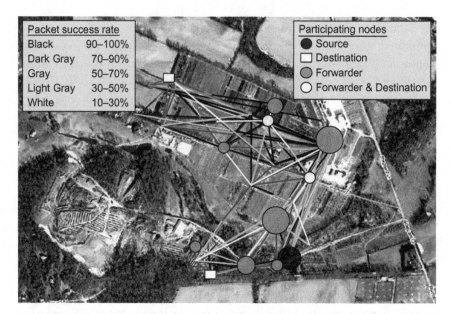

Figure 6.21 Subgraph consists of source nodes (black circles), forwarder nodes (gray circles), forwarder-destination nodes (white circles) and destination nodes (white squares). Size of node (circle) indicates its forwarding rate.

denotes nodes that forward traffic, destinations are white squares and forwarder-destinations are white circles. The area of the circle is proportional to the forwarding factor (pure destinations have been overlaid with an white square for visibility since they do not forward data). The links are colored as in Fig. 6.14. Note that there are nine forwarding nodes including the two destinations that also forward traffic. The subgraph uses many nodes to forward traffic. Efficiency is obtained by having many of these forwarding nodes operate with a small forwarding factor—the rich topology does not come at the expense of having many nodes send at a high rate.

8.2.2 Air Tactical Scenario Results

The results for the airborne-node assisted mobile tactical scenario (Fig. 6.22) were similar to those for the ground tactical scenario. The aircraft helped reduce the number of hops in many source-destination paths. However, due to hidden terminal collisions (most ground nodes can

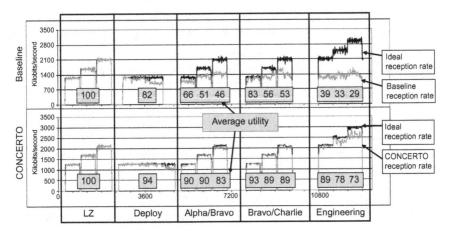

Figure 6.22 Video performance in air scenario.

see the air radios but not each other), there is a high packet loss rate (50 to 70%) from ground to air.

Overall, CONCERTO provided significant gains over the baseline in video traffic delivery. The baseline protocols provided good video only during initial and deployment phases while CONCERTO performed well in all phases of the airborne tactical scenario, even in the presence of many poor quality links (such as those from ground to air). The main reason was the ability of the network coding subgraphs to direct information around the hot-spots (air nodes). It is interesting to note that the air nodes did not help the overall network performance for either the baseline or CON-CERTO algorithms. However, unlike the baseline, CONCERTO was able to provide good performance despite the hot-spots created by the air nodes as its subgraphs provided a rich set of alternative paths.

8.2.3 Tactical Scenario File Transfer Results

File transfer was again an important success for CONCERTO (see Table 6.4). CONCERTO achieved nearly a 100% success rate for file transfers, while the baseline achieved an approximately 20% success rate for file transfers in the tactical multi-hop scenarios. The successful transfer of all files was enabled by the fully-reliable network coding transport protocol

Table 6.4 Average file transfer utility, per phase
File Transfer Results

	Ground		Air	
Phase	CONCERTO	Baseline	CONCERTO	Baseline
Parking lot	100%	100%	100%	100%
Deployment	100%	47%	100%	47%
Alpha/Bravo	100%	26%	100%	16%
Bravo/Charlie	100%	21%	100%	20%

which combines the hop-by-hop repair of a network coded group of packets (generations) with an end-to-end strategy of detecting and recovering entire network coded groups.

9. CONCLUSION AND FUTURE WORK

9.1. Summary

CONCERTO outperformed the baseline in all phases, particularly in the dynamic and challenging phases where video must transit multiple hops over poor quality links. CONCERTO derives its performance benefits from the combination of network coding and the use of a rich subgraph for forwarding. Network coding allows CONCERTO to exploit poor links and to use multiple forwarders to deliver information to mobile nodes. Network coded packets, arriving from different forwarding nodes, have a high probability of finding a path which allows delivery of sufficient information to recover the application data. Thus, CONCERTO achieves a high probability of delivery while efficiently using channel capacity. The baseline protocol suite using flooding has too many nodes transmitting the same packets, and thus reaches capacity at lower application rates.

CONCERTO makes efficient use of the channel capacity by using multiple-paths to the destination and by using link-layer rather than end-to-end retransmissions to overcome link loss. The network coding of new and repair packets allows transmissions to benefit multiple receivers simultaneously. These techniques allow CONCERTO to use scarce bandwidth efficiently and to deliver more data than the baseline protocols. The subgraph-based encoding and multiple-path approach increases the robustness of CONCERTO to the loss of individual links and nodes in a dynamic environment.

9.2. Future Work

While the CONCERTO system has achieved significant gains in packet delivery performance and bandwidth efficiency, it has the potential for further improvement in several directions.

The subgraph construction algorithm has a computational complexity of $O(n^2)$, where n is the number of nodes. We rely on OLSR to discover and propagate complete topology information around the network to enable the routing subgraph computation at the source node of each multicast session. A more scalable alternative can be based on using only local topology information and incremental propagation of routes similar to the distance-vector type of routing algorithms. In a related project (BRAVO, funded by ONR) we are exploring such algorithms based on models of information propagation using Potential Fields [20].

While we believe that CONCERTO performance benefits are independent of the MAC and PHY layers, there may be significant additional performance gains possible by combining network coding with a novel MAC layer protocol and specialized PHY layer hardware that permit multiple packet receptions via multi-user detection [21]. We intend to design such a solution and validate the benefits of combining multiple packet transmissions (via network coding) and multiple packet receptions.

ACKNOWLEDGMENTS

The authors would like to thank the BAE Systems team for their contributions to the CONCERTO project (August 2006–August 2009): Jennifer Costello, Brendan Coyle, Brian DeCleene, Victor Firoiu, Robert Flynn, Sean Griffin, Mark Keaton, Clifton Lin, Greg Lauer, Ryan Metzger, Soumendra Nanda, Tom Porcher, Sean Shen, Joseph Sivak, Charles Tao.

In addition, we would like to acknowledge the contributions of the following institutions: California Institute of Technology, Cornell University, Massachusetts Institute of Technology, Pennsylvania State University, Stow Research, University of Illinois at Urbana-Champaign, and University of Massachusetts at Amherst.

REFERENCES

[1] B. Adamson, C. Bormann, M. Handley, and J. Macker, Negative-Acknowledgement (NACK)-Oriented Reliable Multicast (NORM) Transport Protocol, RFC 5790, November 2009.

[2] Simplified Multicast Forwarding, SMF, IETF draft, draft-ietf-manet-smf-11.

[3] Open Shortest Path First, OSPF, IETF RFC 2328.

[4] T. Clausen, and P. Jaquet, Optimized Link State Routing Protocol (OLSR), IETF RFC 3626.

[5] Protocol Independent Multicast, IETF RFC 4601.

[6] R. Ahlswede, N. Cai, S. Li, and R. Yeung, Network Information Flow, *IEEE Transactions on Information Theory*, 46: 1204–1216, 2000.

[7] S. Li, R. W. Yeung, and N. Cai, Linear network coding. *IEEE Transactions on Information Theory*, 49: 371–381, 2003.

[8] R. Koetter and M. Médard, An Algebraic Approach to Network Coding, *IEEE/ACM Transactions on Networking*, October 2003.

[9] T. Ho, R. Koetter, M. Médard, D. R. Karger, and M. Effros, The Benefits of Coding over Routing in a Randomized Setting, In *Proceedings of 2003 IEEE International Symposium on Information Theory*, June 2003.

[10] D. Lun, M. Médard, R. Koetter, M. Effros, On coding for reliable communication over packet networks. In Proc. 42nd Annual Allerton Conference on Communication, Control, and Computing, Sept.–Oct. 2004, invited paper.

[11] D.S. Lun, Ratnakar, N., Médard, M., Koetter, R., Karger, D.R., Ho, T., Ahmed, E., and Zhao, F., Minimum-cost multicast over coded packet networks, *IEEE Transactions on Information Theory*, Vol. 52, Issue 6, pp. 2608–2623, June 2006.

[12] S. Chachulski, M. Jennings, S. Katti, and D. Katabi, Trading Structure for Randomness in Wireless Opportunistic Routing, SIGCOMM'07, Kyoto, Japan, pp. 169–180.

[13] B. Cain et al., Internet Group Management Protocol, Version 3, IETF RFC 3376, October 2002.

[14] J. Park, D. Lun, Y. Yi, M. Gherla, and M. Médard, CodeCast: A network coding based *ad hoc* multicast protocol, *IEEE Wireless Comm. Mag.*, 2006.

[15] J. Byers, M. Luby, M. Mitzenmacher, and A. Rege, A Digital Fountain Approach to Reliable Distribution of Bulk Data, Proc. of ACM SIGCOMM'98

[16] S. Biswas and R. Morris, ExOR: Opportunistic Multi-hop Routing for Wireless Networks, SIGCOMM'05, Philadelphia, USA, pp. 133–144

[17] Basic Multicast Forwarding, BMF, http://sourceforge.net/projects/olsr-bmf/

[18] Application LAyer REliability System, Alares, https://www.darkcornersoftware.com/confluence/display/ALA/Alares

[19] Nettion, NETwork TestIng and Operational eNvironment, http://www.darkcornersoftware.com/nettion.html

[20] V. Firoiu, and H. Liu, BRAVO: Potential Field-Based Routing and Network Coding for Efficient Wireless MANETs, in preparation.

[21] S. Katti, H. Rahul, W. Hu, D. Katabi, M. Médard, and J. Crowcroft, XORs in the Air: Practical Wireless Network Coding, SIGCOMM'06, Pisa, Italy, pp. 243–254

CHAPTER 7

Secure Network Coding: Bounds and Algorithms for Secret and Reliable Communications

Sidharth Jaggi[1] and **Michael Langberg**[2]

[1]Department of Information Engineering, the Chinese University of Hong Kong, Hong Kong;
[2]Department of Mathematics and Computer Science, the Open University of Israel, Raanana, Israel

Contents

Abstract

Network coding allows network routers to mix the information content in incoming packets before forwarding them. This mixing has been theoretically proven to be very beneficial in improving both throughput and robustness. But what if the network contains malicious nodes? Such nodes may "tap" the network in order to eavesdrop

Network Coding. DOI: 10.1016/B978-0-12-380918-6.00007-X
Copyright © 2012 Elsevier Inc. All rights reserved.

on ongoing communication, and/or may pretend to forward packets originating from the source, while in reality they inject corrupted packets into the information flow so as to disrupt communication.

Since network coding allows routers to mix packets' content, a single corrupted packet can end up corrupting all the information reaching a destination. Unless this problem is solved, network coding may perform much worse than pure forwarding in the presence of such malicious adversaries.

This chapter addresses the task of multicast communication using network coding in the presence of passive eavesdroppers and active jammers. Rather surprisingly, it is shown that high-rate private and reliable communication via schemes that are both computationally efficient and distributed is possible in the settings under study. This chapter summarizes almost a decade of study in the very dynamic and intriguing field of private and reliable network communication. The algorithmic techniques presented cover several paradigms and include tools from the study of combinatorics, linear algebra, cryptography, and coding theory.

Keywords: Network coding, network communication, error correction, wiretap channel, distributed protocols, cryptography, capacity.

1. INTRODUCTION

Network coding allows the routers to mix the information content in packets before forwarding them. This mixing has been theoretically proven to maximize network throughput [1, 29, 40, 44]. For multicast communications it can be done in a distributed manner with low complexity, and is robust to packet losses and network failures [24, 46]. Furthermore, recent implementations of network coding for wired and wireless environments demonstrate its practical benefits [20, 32].

But what if the network contains malicious nodes? Nodes that *tap* the network aim to eavesdrop on ongoing communication. Further, some nodes may pretend to forward packets from source to destination, while in reality they *inject* corrupted packets into the information flow. Since network coding makes the routers mix packets' content, a single corrupted packet can end up corrupting *all* the information reaching a destination. Unless this problem is solved, network coding may perform much worse than pure forwarding in the presence of such malicious adversaries.

This chapter addresses the task of multicast communication using network coding in the presence of passive eavesdroppers and active jammers. Rather surprisingly, it will be shown that high-rate private and reliable communication via schemes that are both computationally efficient and distributed is possible in the settings under study. Despite the complexity introduced by distributed network coding, it turns out that many of the classical results for private and reliable communication over point-to-point links have direct analogs in the network setting. This chapter summarizes almost a decade of study in the very dynamic and intriguing field of private and reliable network communication.

1.1. Overview of Chapter

The chapter consists of four sections. In Section 2 we set the notation, definitions, and model used throughout. In Section 3 we consider communication in the presence of *passive* adversaries who only have eavesdropping capability and wish to learn the information transmitted over the network. The objective in this case is the design of communication schemes that enable *secrecy*, i.e., schemes which do not allow the adversary to learn the information transmitted by the source. In Section 4 we consider *active* adversaries, who have both eavesdropping and jamming capabilities, and whose objective is to cause a decoding error at the terminal nodes. Here, successful communication means *reliable* communication, i.e., correct decoding. In Section 5 we consider again communication in the presence of active adversaries with both eavesdropping and jamming capabilities, however in this case our objective is to design communication schemes that are both reliable and secret. Finally, in Section 6 we briefly note other models that do not fit into the above classifications. We conclude with a discussion in Section 7.

In each of the sections mentioned above, our overview includes between two and three refined models. We first address the *coherent* setting in which the terminal nodes are assumed to know the topology of the network alongside the (realization of the) communication scheme used. The second setting we address is the *non-coherent* setting. Here, no knowledge of the topology and/or code being used is assumed to be present at the terminal nodes. In both settings above, we follow an *information-theoretic* analysis assuming that the adversary has unlimited computational power, and has full knowledge of the network topology and the communication scheme in use. We stress that any achievable rate R in the non-coherent setting is also achievable in the coherent one, and any upper bounds presented on R for the coherent setting also hold in the non-coherent one. Finally, we also consider the case in which the adversary is computationally limited, and discuss schemes with are conditioned on certain *cryptographic* assumptions. We present an extended discussion in Section 4, while cryptographic schemes for Sections 3 and 5 can be reduced to our discussion in Section 4. Each of the refined models is presented in more detail in Section 2.

2. MODEL

We use a general model that encompasses both wired and wireless networks. To simplify notation, we consider only the problem of communicating from a single source to a single destination. But similarly to most network coding algorithms, our techniques generalize to multicast traffic.

2.1. Threat Model

There is a source, Alice, who communicates over a wired or wireless network to a receiver Bob. There is also an attacker Calvin, hidden somewhere in the network. Calvin aims to prevent or minimize the transfer of information from Alice to Bob, and/or to eavesdrop on it. He can observe some or all of the transmissions, and can inject his own. When he injects his own data, he pretends it is part of the information flow from Alice to Bob.

Calvin is quite strong. In both the coherent and non-coherent setting we assume that Calvin knows the encoding and decoding schemes of Alice and Bob, and the network code implemented by the interior nodes. He also knows the exact network realization. The computational power of Calvin is assumed to be unbounded unless specifically mentioned otherwise. In the latter case we will specify the exact computational problems limiting Calvin (e.g., the Discrete-Log problem).

2.2. Network and Code Model

Network Model

The network is modeled as a graph, which is acyclic except when specified otherwise. Each transmission carries a packet of data over an edge directed from the transmitting node to the observer node. The graph model captures wired networks. For wireless networks, one may assume a model in which the network is a *hypergraph* in which each edge is determined by instantaneous channel realizations (packets may be lost due to fading or collisions) and connects the transmitter to all nodes that hear the transmission. In this survey we will focus on the wired setting, although several of the results extend naturally to the wireless setting as well. Throughout, the graph is unknown to Alice and Bob prior to transmission in the non-coherent setting, but is assumed to be known in the coherent setting.

Source
Alice generates incompressible data that she wishes to deliver to Bob over the network. To do so, Alice encodes her data as dictated by the encoding algorithm (described in subsequent sections).

Adversary
Calvin is assumed to control certain links of the network. We assume that Calvin can corrupt the information transmitted on any subset of z_O links of the network and can observe the information on z_I links. The set of links controlled by Calvin is unknown to both Alice and Bob. Moreover, this set of links does not undergo any changes throughout the entire block-length of communication. In Section 3 we assume that z_I is unlimited but $z_O = 0$ (here, the objective is to design secret communication schemes). In Sections 4 and 5 we assume a positive value for z_O and discuss different settings of z_I (here, the objective is to respectively design reliable communication schemes, and schemes that are both reliable and secret).

In our model the error imposed by the Byzantine adversary Calvin is assumed to be *added* to the original information transmitted on the network. One can also consider a model in which these errors *overwrite* the existing information transmitted by Alice. We stress that if Calvin is aware of transmissions on links, these two models are equivalent. Overwriting a message x with z is equivalent to adding $-x + z$ over the field over which coding is performed. However, when Calvin is unaware of the information transmitted over the links, these models may differ.

Receiver
The receiver Bob decodes his incoming information using decoding procedures that are discussed in subsequent sections.

Network Transform
In many of the proposed schemes, the network performs classical distributed network coding [24]. Specifically, each packet transmitted by an internal node is a random linear combination of its incoming packets.

Thus, the effect of the network at the destination can be summarized by $Y = TX + T'Z$, where the matrix X represents the encoded source information, the matrix Z represents the error specified by the network, the matrix Y corresponds to the incoming information at the terminal node, and both T and T' represent the linear transforms resulting from the network coding scheme. As is common in the network coding literature, one assumes that the coding is done over a certain finite field $\mathbb{F}$.

Definitions

We define the following concepts. The *network capacity*, denoted by C, is the time-average of the maximum number of packets that can be delivered from Alice to Bob, assuming no adversarial interference, i.e., the max flow. It can be also expressed as *the min-cut from source to destination*. (For the corresponding multicast case, C is defined as the minimum of the min-cuts over all destinations.) The *error probability* is the probability that Bob's reconstruction of Alice's information is inaccurate. The rate R is the number of *information* symbols that can be delivered on average, per time step, from Alice to Bob. Rate R is said to be achievable if for any $\epsilon_1 > 0$ and $\epsilon_2 > 0$ there exists a coding scheme of block length n with rate $\geq R - \epsilon_2$ and error probability $\leq \epsilon_1$. The *capacity* of a certain adversarial setting is the maximum achievable rate under this setting.

3. EAVESDROPPING SECURITY

We start by considering private communication in the context of multicast network coding.

3.1. The Coherent Case

We consider the rate of secret communication (also referred to as *secure* communication) in the coherent setting in the presence of a hidden eavesdropper that has access to z_I links of the network. Specifically, we denote the information transmitted by the source as the random variable X, that received by terminal t_i as Y_i, and that observed by the adversarial eavesdropper as Z. Analogously to the conditions Shannon [64] used to define *perfectly secure* systems, we require that $H(X|Y_i) = 0$ (implying that the terminal is able to deduce the source information) and $I(X; Z) = 0$ (implying that the communication is secure). Letting C denote the maximum multicast communication rate (in the absence of an eavesdropper), we show that one can securely communicate at *optimal* rate $R = C - z_I$. This rate is the best possible, as one cannot communicate securely at a higher rate even over the *one-hop* unicast network consisting of a single source s that wishes to communicate with a single terminal t over C multiple (s, t) links. This is implied by the following standard argument. Let $\bar{Z}$ denote the random variable corresponding to the information on the links not observed by the eavesdropper (hence $Y_i = (Z, \bar{Z})$). Then:

$$R = H(X) = H(X|Y_i) + I(X; Y_i) = H(X|Y_i) + I(X; Z, \bar{Z}) \qquad (1)$$

$$= I(X; Z, \bar{Z}) = I(X; Z) + I(X; \bar{Z}|Z) \qquad (2)$$

$$= I(X; \bar{Z}|Z) \leq H(\bar{Z}) \leq C - z_I. \qquad (3)$$

Equalities (1) $\Rightarrow$ (2) and (2) $\Rightarrow$ (3) follow from our conditions for secure communications, and the remaining (in)equalities from standard information (in)equalities and the fact that $\bar{Z}$ has at most $C - z_I$ links.

Secure communication in the context of coherent network coding has been addressed in several works over the last decade. This line of study was initiated by Cai and Yeung in [9] where they consider enhancing any linear network coding scheme which allows communication at rate C (in the absence of an eavesdropper) to one which is secure. Enhancing an existing linear network coding scheme (such as that of Jaggi *et al.* [29]) is done in an *end-to-end* manner. Namely, internal nodes of the network are oblivious to the fact that communication is done in the presence of an adversarial eavesdropper, and follow the original coding scheme. The presence of an adversarial eavesdropper is dealt with by an enhanced encoding at the source node and by specialized decoding at terminal nodes. To this end, the enhanced encoding of the source includes two steps. Primarily the source takes its $C - z_I$ characters of information X and appends to it a uniformly distributed random vector R of z_I characters to obtain (X, R). Secondly, (X, R) over goes a certain invertible linear transform T resulting in the *message* M of length C. The message M is now transmitted over the network using the original (perhaps insecure) network coding scheme. On decoding, a terminal first recovers M and then via T recovers X.

We now address the security of the scheme suggested in [9]. Considering the original coding scheme as fixed, the code design of [9] involves specifying the matrix T which in turn defines the message $M = T(X, R)$ to be transmitted. Specifically, one needs to design the matrix T such that any z_I linear combinations of M resulting from the linear encoding of the original network coding scheme do not reveal information on the value of X. This is a non-trivial task and the construction of T in [9] is done in a greedy iterative manner that resembles the Gilbert construction for error correcting codes [18]. The analysis of [9] uses a field $\mathbb{F}$ of size $q \geq \binom{|E|}{z_I}$, which is exponential in network parameters. (This bound on q follows from the fact that there are at most $\binom{|E|}{z_I}$ different possible subsets of z_I links that the adversary may eavesdrop on, and the matrix T has to be resilient to each scenario.) An extension to *imperfect security* is also addressed in [9]. Using essentially the same construction and proof technique, it is shown that allowing $I(X; Z) \leq i$ (instead of $I(X; Z) = 0$) one can obtain rate $C + i - z_I$ (here one assumes $i \leq z_I$).

Theorem 1 ([9]). *Let G be an acyclic network with cut capacity C. Then, the coherent secure capacity in the presence of an adversary that may eavesdrop on up to z_I of the links of G is $C - z_I$.*

Following the paradigm of [9], Feldman *et al.* [14], study the achievable trade-off between rate and field size q. Namely, referring to the matrix T as a "filtered secret sharing scheme", they show that finding T is equivalent to finding an error-correcting code with certain generalized distance properties. The latter is obtained via a random linear construction similar to that of Varshamov [79]. Using this connection, Feldman *et al.* [14] show that for any $\sigma > 1$ one can efficiently construct a matrix T that allows secure communication at rate $C - \sigma z_I$ and field size $q = |E|^{\Omega(\frac{1}{\sigma-1})}$. The intuition is that any scheme with the appropriate generalized distance has the property that projecting the linear transform onto z_I links results in a uniformly distributed random variable from the eavesdropper's perspective.

In [61, 62], Rouayheb *et al.* study the two-step paradigm of [9] and take a different approach in which they concentrate on the design of the internal network coding scheme instead of the design of T. Namely, [61, 62] tie the task of coherent secure communication with that of *the wiretap channel of type II* introduced in the seminal work by Ozarow and Wyner [57, 58]. In the latter, coset coding is used to enable secure communication at rate $C - z_I$ over the simple one-hop network mentioned previously. The authors of [61, 62] observe that the naïve approach—one in which the source pre-encodes its information using the coset coding techniques of [57, 58] (via T) and then performs any feasible network coding scheme—does not necessarily yield secure communication. However, [61, 62] show that if the network coding scheme satisfies certain requirements with respect to the specific coset coding used, then a secure rate of $C - z_I$ is indeed obtained. To obtain such network codes that have a good fit with a given pre-encoding scheme T, Rouayheb et al. modify the deterministic algorithm of [29] for the construction of network coding schemes (in the absence of an eavesdropper) and obtain efficiently constructible secure codes at optimal rate $C - z_I$ with field size roughly $|E|^{z_I}$. To reduce the field size to one which is independent of the size of the graph G and only depends on C, z_I and the number of terminals in the multicast connection t, [61, 62] use ideas from [42, 43] and obtain a field-size of $\binom{2(C-z_I)^3 t^2}{z_I - 1} + t$, which is independent of $|V|$ and $|E|$ but still exponential in other network parameters.

Pre-encoding using coset coding is further investigated by Ngai et al. [53] in which a comprehensive study is performed. Motivated by the work of Wei on generalized Hamming weight for linear block codes [83], Ngai et al. define the notion of "Network Generalized Hamming Weight" and "Network MDS" codes. Roughly speaking, these notions tie block error-correcting codes with network coding schemes and suffice to characterize pre-encoding schemes T that allow secure communication when combined with a given network coding scheme (and *vice versa*).

Considering a *weak* notion of security Bhattad et al. [7] study the scenario in which the eavesdropper may indeed obtain partial information regarding the messages multicast over the network, however this partial information does not suffice to deduce the exact value of any of the characters of the source information X. For example, on transmission of a message with two symbols a and b, eavesdropping on the sum $a + b$ reveals partial information about the message (a, b) but does not reveal the exact value of either a or b. Similarly to the paradigm of [9], the work of [7] shows that any network coding scheme of capacity C can be turned into a weakly secure one by multiplying the source information X with a certain matrix T. For this reduction to work, [7] requires a field of size roughly $|E|^{\frac{k}{C-k}}$, where $k \le z_I$ is a parameter corresponding to the amount of information that the eavesdropper may obtain on the links under his control. Among other related questions, [7] also addresses the natural question of perfect/weak security of a random linear network code without any pre-encoding via T. Here, as in [26], a random linear network code is one in which the linear coefficients governing the coding scheme are all chosen uniformly and independently at random from the underlying field $\mathbb{F}$. In addition one assumes that the actions of the eavesdropper (namely which links to control) are independent of these random choices. In this setting a trade-off between field size and probability of error is given (for both perfect and weak security). Roughly speaking, if one allows an ϵ probability of error in the design process, then the field size (when compared to the bounds of [9] for standard security and [7] for weak security) are to be multiplied by a factor of $1/\epsilon$.

The works mentioned above all focused on acyclic networks. Following the analytical techniques of [9], Jain [30] studies secure network coding in the general (not necessarily acyclic) setting. Namely, in [9] a general analysis of secure communication was conducted in the case where the eavesdropper may choose a set of links $A \subseteq E$ from a given set system $\mathcal{A}$. When the adversarial eavesdropper may control at most z_I links, then $\mathcal{A}$ is just the set system consisting of all subsets of at most z_I links. This general analysis is strengthened in [30] for the cyclic case in which there is a single source node, a single terminal node and one wishes to communicate at unit rate. Namely, necessary and sufficient conditions for secure unit rate communication are presented in terms of the topology of G and the set system $\mathcal{A}$. In a nutshell, after preprocessing the graph G and removing from G nodes from which information cannot reach the single terminal node, the necessary and sufficient condition for secure communication is the existence of a single *untapped* path, i.e., a path not seen by the adversary, from sender to receiver when considering the preprocessed graph as undirected. A characterization for higher communication rate or more terminal nodes is an open problem.

3.2. The Non-Coherent Case

In this section we focus on communication in *non-coherent* settings. That is, the network topology and network coding operations are unknown in advance to the communicating parties. Despite this restriction, it can be shown that essentially the same performance as in the coherent setting can still be achieved. We focus on two lines of work—schemes with randomized source encoding functions, and those with deterministic source encoding functions.

The construction of Feldman *et al.* [14] mentioned in the previous section falls in the former case. In their construction, the linear filter that the source node passes its message through is obtained by randomly choosing a matrix of the appropriate dimension. Interior nodes in the network perform random linear combinations over sufficiently large finite fields, in the spirit of the distributed random linear network coding scheme of Ho *et al.* [26]—indeed, they can be unified to generate a distributed non-coherent random linear network code that is perfectly secure against a wiretapper that can eavsdrop on at most z_I links. They demonstrate that with high probability over these random choices, the generated linear code is perfectly secure against eavesdropping by any adversary that can wiretap at most z_I links. Further, if the min-cut of the network is denoted by C, the rate at which communication can be carried out in this scheme can be made arbitrarily close to $C - z_I$ as the field size over which the scheme is designed grows without bound. They also show that the field size required for such a scheme can be much smaller in general, than if one required secret communication at a rate exactly equaling $C - z_I$.

Theorem 2 ([14]). *For any $\sigma > 1$, for any field size $q \geq \max\{|E|^{\Omega(1/(\sigma-1))},\ |S|\}$, (where S denotes the set of terminals), there exists a feasible linear network coding scheme with rate $C - \sigma z_I$ which is perfectly secure against a wiretapper eavesdropping on z_I links. Further, a random choice of source filter achieves this performance with high probability.*

In contrast, they also show that if the desired rate of communication is exactly $C - z_I$, then the minimum field size required is at least $|E|^{\Omega(\sqrt{z_I/\log z_I})}$. The reason that a small gap from the capacity results in a significant reduction of field size is that the number of linear transforms that take a message at rate $C - \sigma z_I$ to a message of dimension C is significantly larger than the number of linear transforms that take a message at rate $C - z_I$ to a message of dimension C.

Silva and Kschischang [78] consider a deterministic source encoder that can be overlaid onto a non-coherent random linear network code (for instance, that of Ho *et al.* [26]). Their work is motivated by Rouayheb and Soljanin's formulation of a wiretap network and builds on their results. They propose a coset coding scheme based on "maximum rank-distance" (MRD) codes, that neither imposes any constraints on, nor requires any knowledge of, the underlying network code. In other words, for any linear network code that is feasible for multicast, secure communication at the maximum possible rate is achieved with a fixed outer code. In particular, the field size can be chosen as the minimum required for multicasting. The essence of their approach is to use a vector linear outer code over a block length n that is, in fact, also a linear code over the extension field $\mathbb{F}_{q^n}$.

Theorem 3 ([78]). *A perfectly secure communication scheme at rate $C - z_I$ can be achieved by using any feasible $\mathbb{F}_q$-linear network code in conjunction with a fixed end-to-end coset coding scheme based on any linear MRD (C, z_I) code over $\mathbb{F}_{q^n}$.*

4. JAMMING SECURITY

In this section we consider the design of network codes that enable reliable error-detection and communication in the presence of active jammers that have both eavesdropping and jamming capabilities. Our discussion follows the outline presented in the Introduction.

4.1. The Coherent Case

For the problem of error correction we first study the rate of reliable communication in the coherent setting in the presence of an active jammer that can jam z_O links of the network and observe all links of the network. In a nutshell, we show that this rate is $C - 2z_O$ for $C \geq 2z_O$ and 0 otherwise. Namely, the rate is equal to $(C - 2z_O)^+$. We start by considering the class of one-hop unicast networks. In a one-hop unicast network there is a single source s that wishes to communicate with a single terminal t over C multiple (s, t) links. We assume that the links may carry a single character from a given alphabet Σ of size q, and that the source wishes to transmit R characters of Σ to t. It is not hard to verify that the task of designing a communication scheme with rate R that allows reliable communication over one-hop unicast networks in the presence of an adversary that may jam z_O of the links is equivalent to the design of $[C, R]$ error correcting codes that are resilient to z_O errors (i.e., have minimum distance $2z_O + 1$).[1]

---

[1] By a similar argument it can also be observed that errors injected by an adversary who can jam z_O links may be detected if, and only if, the minimum distance of the code is at least $z_O + 1$, i.e., the maximum rate at which adversarial errors can still be detected is $C - z_O$. Here, it is crucial that we assume the coherent setting in which the (network) code is known to all parties. If, however, the adversary cannot observe everything in the network, the work of [25] demonstrates that errors can still be detected for any rate of communication strictly less than C.

Let q be the size of Σ. There are multiple bounds on the rate $R(C, z_O, q)$ of error correcting codes over alphabets of size q with block length C and minimum distance $2z_O + 1$. It is well-known that $R(C, z_O, q) \leq C - \Delta$ for $\Delta = \log_q \left(\sum_{i=0}^{z_O} \binom{C}{i}(q-1)^i \right)$, e.g., [48]. This bound is referred to as the *sphere packing* or *Hamming* bound, and follows from a simple volume argument. As q approaches infinity it can be verified that this bound approaches $C - z_O$. This bound holds for all types of errors—random *or* adversarial. Further, the *Singleton bound* (e.g., [48]), derived using the pigeonhole principle, shows that $R(C, z_O, q) \leq C - 2z_O$. Hence for sufficiently large alphabet sizes q, the Singleton bound is tighter than the sphere packing bound.

What about lower bounds on $R(C, z_O, q)$? Several coding techniques [48] (including for example Read-Solomon codes) imply C-block error correcting codes resilient to z_O errors whose rate equals $C - 2z_O$. Most relevant to this chapter are the works of Gilbert [18] and Varshamov [79] that show that $R(C, z_O, q) \geq C - \Delta$ where $\Delta = \log_q \left(\sum_{i=0}^{2z_O} \binom{C}{i}(q-1)^i \right)$. Notice that the summation in this case is from 0 to $2z_O$ (as apposed to z_O in the Hamming bound). The discussion above implies that as q tends to infinity, the Singleton bound is tight and corresponds to the capacity of one-hop unicast networks in the presence of jammers.

A natural and intriguing question is whether the above setting also holds in more complicated networks as well. This question was studied by Cai and Yeung in [10, 87] and was answered in the affirmative. Namely, [10, 87] show an analog to the Hamming bound, Singleton bound, and Gilbert-Varshamov bound in the coherent network coding setting. Moreover, they show their Singleton-type bound for networks equals their Gilbert-Varshamov type bound for large values of q.[2] The crux of their analysis lies in understanding the combinatorial nature of information transmitted on minimum *cut-sets* of the network that separate source terminal pairs.[3] In what follows we give an overview of the results in [10, 87].

---

[2] Tighter bounds on the field-size required were obtained by [2] and in [3] the authors demonstrated that the field-size requirement can be drastically reduced if one reduces the required rate slightly—the result is analogous to the one obtained by Feldman *et al.* [14] for eavesdropping security. High-complexity algorithms for adversarial network error-correction were also obtained in [88].

[3] The work of [74] translates this analysis (and further results in [10, 87] to be presented shortly) into the language of "matrix channels", as discussed in Section 4.2.

Consider any given network $G = (V, E)$ with (error free) capacity C. Let A and B be a partition of V, and let cut(A, B) denote the set of links directed from a node in A to a node in B. To obtain an analog to the Hamming bound for networks, [87] considers the information transmitted over cut-sets cut(A, B), or to be precise, the mapping between the source information X and the information $Z^m = Z_1, \ldots, Z_m$ transmitted over the cut-set. Here $m = |\text{cut}(A, B)|$. Roughly speaking, if there are no links directed from B to A in G, it must be the case that Z^m is an $[m, 2z_O + 1]$ error correcting code. This follows directly by the fact that decoding at terminal t is solely a function of Z^m. Indeed, if Z^m did not have minimum distance $2z_O + 1$ then a malicious jammer corrupting z_O links from cut(A, B) may cause a decoding error at t. Note that the reduction above relies on the lack of edges from B to A, otherwise errors on certain links of cut(A, B) may affect other links in cut(A, B) (such effects do not occur in the standard model of error correcting codes). Once the reduction between network communication and error correcting codes is established, the Hamming-type bound and Singleton-type bound follow.

Theorem 4 (Network Hamming Bound). *Let G be an acyclic network with (error free) cut capacity C, in which each link can carry a single character of an alphabet Σ of size q. Then the coherent capacity when at most z_O of the links of G are jammed is at most $C - \Delta$ where:*

$$\Delta = \log_q \left(\sum_{i=0}^{z_O} \binom{C}{i} (q-1)^i \right).$$

As the field size q approaches ∞ with fixed C and z_O, this bound approaches $C - z_O$.

As for classical error correcting codes, a stronger bound for the network adversarial error case for large q is the network analog of the Singleton bound.

Theorem 5 (Network Singleton Bound). *Let G be an acyclic network with (error-free) cut capacity C. Then, the coherent capacity in the presence of an adversary that may jam up to z_O of the links of G is at most $(C - 2z_O)^+$.*

We now turn to discuss lower bounds on the coherent capacity in the presence of an adversary that may jam up to z_O links. In [10] a Gilbert-Varshamov bound in the context of network communication is derived. It is well-known that in the error-free coherent setting, one can communicate the set Σ^C of distinct messages successfully over the network using, for example, linear network codes that are constructed at random. Using such network codes, the main idea in [10] is to carefully construct a subset of messages $W \subset \Sigma^C$ with the property that *no matter which error pattern* is chosen by the adversary, each terminal is able to correctly distinguish the message $w \in W$ transmitted. Namely, two words x and x' of Σ^C are said to be non-separable if there exist two error patterns e and e' such that the information reaching a terminal node when x is transmitted and the adversary applies the error pattern e is *identical* to that received when x' is transmitted and e' applied. The objective in [10] involves identifying a *large* subset W for which each $w \neq w' \in W$ are separable. The crux of their analysis lies in a careful study, for a given $x \in \Sigma^C$, of the subset V_x of possible words x' such that x and x' are non-separable. Bounding the size V of V_x and following the greedy technique of Gilbert [18] will yield sets W of size q^C/V. Moreover, using a Varshamov-type approach one is able to bound V by q^{2z_O} and obtain a linear W of size q^{C-2z_O}.

Theorem 6 (Network Gilbert-Varshamov Bound). *Let G be an acyclic network with (error free) cut capacity C in which each link can carry a single character of an alphabet Σ of size q. If q is sufficiently large, then the coherent capacity in the presence of an adversary that may jam up to z_O of the links of G is at least $(C - 2z_O)^+$.*

Corollary 1 (Coherent Capacity). *Let G be an acyclic network with (error free) cut capacity C. Then, the coherent capacity in the presence of an adversary that may jam up to z_O of the links of G is $(C - 2z_O)^+$.*

4.2. The Non-Coherent Case

We now consider the rate of reliable communication in the *non*-coherent setting in the presence of a hidden active jammer that can jam z_O links of the network. In this setting neither the network topology nor the network code are known in advance. We show that even then, the same rate of $(C - 2z_O)^+$ is achievable as in the coherent case. In fact, interior nodes in the network can be oblivious to the presence of adversaries, and may just perform any "good" predesigned network coding operations (such as deterministic multicast network coding, or distributed random network coding). All the complexity is absorbed into the encoder and decoder, which nonetheless have computational complexity that is polynomial in network parameters.

The key to such performance lies in the following observations. As noted in Section 2, if the network performs linear network coding, the relationship between the source's information X, the fake information Z injected by the adversary, and the information received by the receiver can be expressed as:

$$Y = TX + T'Z. \tag{4}$$

This relationship between X and Y is denoted as the (linear) *operator channel*.

The work of [38, 39] contained the following insights. Let $\mathcal{X}$, $\mathcal{Y}$, and $\mathcal{Z}$ denote the row-spaces of the matrices X, Y, and Z. Then Equation (4) implies that the vector-space $\mathcal{Y}$ is just the direct sum of the vector-spaces $\mathcal{X}$ and $\mathcal{Z}$, i.e., the smallest vector-space containing both $\mathcal{X}$ and $\mathcal{Z}$. They then noted that a *subspace metric* $d_S(.,.)$ can be defined on the set of all subspaces of $\mathbb{F}_q^n$. This is as follows—for any subspaces $\mathcal{U}$ and $\mathcal{V}$ of $\mathbb{F}_q^n$:

$$d_S(\mathcal{U}, \mathcal{V}) = \dim \mathcal{U} + \dim \mathcal{V} - 2 \dim (\mathcal{U} \cap \mathcal{V}).$$

It can be seen that this definition does induce a metric—in particular, the triangle inequality is satisfied by $d_S(.,.)$.

This then indicates a strategy for "good" code design for the operator channel, closely paralleling classical algebraic code designs (such as Reed-Solomon codes). The communicating parties choose in advance a codebook comprising of subspaces of $\mathbb{F}_q^n$, such that each pair of subspaces have a subspace distance of at least $4z_O + 1$ between them.[4] In [38, 39] the authors demonstrate that it is possible to choose such a codebook with at

---

[4] The reason that the appropriate choice is $4z_O + 1$ rather than the more "intuitive" $2z_O + 1$ one would expect from classical coding theory is as follows. Each packet injected by the adversary may, in the worst case, reduce the dimension of the row-space of TX by one, and simultaneously add a vector to it that is in the row-space of TX' for some $X' \neq X$. Hence each packet injected by the adversary can change the subspace distance by up to two. An alternative metric, the *injection metric* defined in [74] does not require this extra factor of two.

least $q^{(C-2z_O)(n-C)}$ elements. For a sufficiently large field size q and packet length n, this approaches q^{C-2z_O}.

The decoder then does the following—it finds the codeword in the codebook that is closest in subspace distance to the observed space $\mathcal{Y}$. Since the adversary controls at most z_O links, the dimension of $\mathcal{Z}$ is at most z_O, and hence this decoding algorithm is guaranteed to work correctly.

The authors of [38, 39] demonstrate computationally efficient encoding and decoding of such codes via codes based on *linearized polynomials*, which are analog of Reed-Solomon codes from classical algebraic coding theory. They then demonstrate in [71, 73, 77] alternative decoding methods of such codes by using rank-metric decoding algorithms proposed by Gabidulin [17].

Taken together, these results imply the following elegant theorem.

Theorem 7 (Noncoherent Capacity for Adversarial Errors). *The noncoherent capacity in the presence of an adversary that may jam up to z_O of the links of G is $(C - 2z_O)^+$. This can be achieved by codes that have computational complexity $\mathcal{O}(C^2 n)$.*

The problem of *detecting* (rather than correcting) adversarial network errors in a non-coherent setting is considerably more straightforward. In a scheme proposed in [25] the source appends a non-linear hash to each packet of the data contained within it. They then show that as long as there is even one uncorrupted path from the source to the destination, then arbitrary errors by the adversary can be detected with high probability, via a low-complexity scheme.

Related work also considers the case of random errors on links rather than adversarial errors. In this model the matrices Z are chosen uniformly at random from the set of $z_O \times n$ matrices, rather than deliberately chosen by an adversary so as to minimize the rate at which the sender and the receiver can communicate with each other. Hence, one could in principle hope for a higher rate than with adversarial errors. The work in [50] and subsequently the work in [76] show that this is indeed the case. The proof in [76] is admirable in its succinctness, and we sketch the main ideas here.

By assumption (or with high probability under random network code design as in [26]), the transfer matrix T may be assumed to be invertible.

Hence (4) may be rewritten as:

$$Y = T(X + T^{-1}T'Z). \tag{5}$$

Here $T^{-1}T'Z$ may also be assumed to be uniformly distributed over the set of all matrices of that dimension.[5] The code construction is then very similar to the random code construction in [26], except that X is padded with rows and columns comprising entirely of zeroes. More precisely, the first z_O rows and columns of X are all set to be zero, and the remaining $(C - z_O) \times (n - z_O)$ sub-matrix comprises of a $(C - z_O) \times (C - z_O)$ identity sub-matrix, and a $(C - z_O) \times (n - C)$ payload matrix U, as in:

$$\begin{bmatrix} 0_{z_O \times z_O} & 0_{C \times z_O} & 0_{n \times C} \\ 0_{(C-z_O) \times z_O} & I_{(C-z_O) \times (C-z_O)} & U \end{bmatrix} \tag{6}$$

Then it can be shown [76] that if the top left $z_O \times z_O$ sub-matrix of Y is full rank, then by Gauss-Jordan elimination the row-reduced form of the received matrix Y equals:

$$\begin{bmatrix} Z_1 & 0_{C \times z_O} & Z_2 \\ 0_{(C-z_O) \times z_O} & TI_{(C-z_O) \times (C-z_O)} & TU \end{bmatrix} \tag{7}$$

But if this is the case, since T is assumed to be invertible, the information payload U can be reconstructed from the last $C - z_O$ rows of (7). The only remaining step is to demonstrate that Y does indeed satisfy, with high probability over Z, the rank constraint assumed above. It is shown in [76] that for large q or n this probability is at least $1 - o(1/q^{n-2z_O})$.

This leads to the following theorem.[6]

Theorem 8 (Noncoherent Capacity for Random Errors). *Let G be an acyclic network with (error free) cut capacity C. Then with probability at least $1 - o(1/q^{n-2z_O})$ the non-coherent capacity in the presence of random packets injected on at most z_O of the links of G is $C - z_O$. This can be achieved by codes that have decoding complexity $\mathcal{O}(C^2 n)$.*

---

[5] This turns out to be the worst case—if $T^{-1}T'Z$ is not uniformly distributed due to rank deficiency in $T^{-1}T'$, the problem may be transformed linearly into another one with different parameters where in fact this is the case.

[6] A similar result and algorithm for random errors was also independently proposed in [88].

Lastly, we touch upon an alternate schema for efficient non-coherent network error correction, proposed in [27, 28] in parallel to the work in [38, 39]. While the rates achievable are asymptotically equivalent in the limit of large field size q and packet length n, the parameters for [27, 28] are generally inferior, in that the required q and n are larger in [27, 28], and the computational complexity is $\Theta(n^3)$ rather than $\mathcal{O}(C^2 n)$ as in [38, 39].

However, one advantage of the proof techniques in [27, 28] is that they allow for computationally efficient "linear list-decoding". A code is said to be *l-list decodable* at rate $R(C, z_O, q)$ if, given the constraint z_O on the set of error patterns, the decoder can always output a list of size at most l which is guaranteed to contain the transmitted codeword. Further, it is said to be *linear* list-decodable at that rate if the list can be represented in the form of an affine shift of a subspace of $\mathbb{F}_q^n$. That is, every vector in the list is of the form $\mathbf{v} + \mathcal{L}$, for some fixed vector $\mathbf{v}$ and some fixed subspace $\mathcal{L}$ with l elements. Then:

Theorem 9 (Linear List-Decoding). *There exist codes of rate $C - z_O$ that are linear q^{C^2}-list decodable in the presence of an adversary that may jam up to z_O of the links of G. The computational complexity of such codes is $\mathcal{O}(C^3 n)$.*

The idea is as follows. The encoder chooses a codebook comprising of q^{C-z_O} matrices X, each of rank $C - z_O$. Since the rank of Z is at most z_O, therefore the rank of the matrix whose rows comprise the rows respectively of X and Z is at most C—without loss of generality, we henceforth assume that it is in fact exactly C (if not, similar arguments hold for smaller values of the rank).

The decoder selects C linearly independent columns of Y, and denotes the corresponding matrix Y^s. The columns of X and Z corresponding to those in Y^s are denoted X^s and Z^s respectively. By Equation (4), $Y^s = [T|T'] \begin{bmatrix} X^s \\ Z^s \end{bmatrix}$. Also, since Y^s acts as a basis for the columns of Y, we can write $Y = Y^s F$ for some matrix F. The decoder can compute F as $(Y^s)^{-1} Y$. Therefore Y can also be written as:

$$Y = [T|T'] \begin{bmatrix} X^s F \\ Z^s F \end{bmatrix}. \tag{8}$$

Comparing Equations (4) and (8), and again using the assumption that $[T|T']$ is invertible (with high probability) implies that:

$$X = X^s F, \tag{9}$$

$$Z = Z^s F. \tag{10}$$

In particular, Equation (9) gives a linear relationship on X that can be leveraged into a list-decoding scheme for the decoder. The number of variables in X^s is C^2. Therefore the entries of the matrix X^s span a vector space of dimension C^2 over $\mathbb{F}_q$. Bob's list is the corresponding C^2-dimensional vector space $\mathcal{L}$ spanned by $X^s F$.

Such a list-decoding result is useful in a variety of settings. For instance, in [27, 28] this result is used as the first stage of a non-coherent network error correcting code—first this result is used by the decoder to generate an affine subspace containing the source's message X, and then the decoder refines this list using extra constraints imposed on the codebook as part of code design. Even though the size of this list is large (q^{C^2}) this refinement procedure can be done computationally efficiently, since the list is affine. Another use of this list decoding result is seen in the next section, on cryptographic protocols.

Note: A special class of errors is that of "packet erasures". As has been observed by several authors (see for instance [38]), z_O packet erasures, whether random or adversarial, correspond to a rate-loss of at most z_O, in contrast to a rate-loss of $2z_O$ in the presence of adversarial errors. Hence the best achievable rate in the presence of z_O erasures is $C - z_O$.

4.3. The Cryptographic Setting

In this section we address adversarial jammers that are computationally bounded. Namely, jammers against which one can apply certain cryptographic primitives. In this line of study, one assumes that certain computational tasks (such as Discrete Log or Factoring) are intractable, and based on these assumptions design a feasible communication scheme. We show the ability to communicate at rate $C - z_O$ in the presence of a computationally bounded adversary that can corrupt up to z_O links of the network. Notice that this improves on the rate of $C - 2z_O$ presented in Sections 4.1 and 4.2 in which the jammer has no computational limitations. Also notice that this rate is the best possible. Since most of the schemes described below proceed by first detecting adversarial attacks and then discarding erroneous packets, network error detection is a direct by-product of the schemes.

Roughly speaking, the works we survey have one of two flavors: *in-network* authentication or *end-to-end* authentication. In the in-network setting, one designs certain authentication mechanisms that allow internal nodes of the network to identify information packets that have been corrupted by the jammer. Once such faulty packets are found, the internal nodes of the network may discard them. This reduces communication in the presence of a jammer to that in which the jammer is absent—but some links of the network are not able to transmit information. The latter scenario, for which standard random linear network coding schemes (e.g., [26]) allow reliable communication, is well understood. There are several challenges in this line of study. These include the design of efficient signature schemes that are on one hand closed under linear coding operations (such signature schemes are referred to as *homomorphic* [4–6, 12, 31, 49, 60]) and on the other do not need an elaborate infrastructure to support key distribution among internal nodes of the network. In-network authentication indeed guarantees communication at rate $C - z_O$, however, in many cases a higher rate is achievable (depending on the exact links controlled by the jammer).

In end-to-end authentication, internal nodes of the network are oblivious to the fact that communication is done in the presence of an adversarial jammer, and follow standard coding protocols used commonly when a jammer is absent. The presence of the jammer is dealt with by an enhanced encoding at the source node and by specialized decoding at terminal nodes. End-to-end schemes have obvious advantages in code management over in-network authentication, and as in-network schemes they promise rate $C - z_O$. However, when compared to in-network authentication on an "instance to instance" basis it may be true that end-to-end authentication obtains a lower rate (here the location of the jammer comes into play—end-to-end authentication schemes assume that the adversary locates itself in a worst-case manner in the network, and hence might be unduly pessimistic).

In-Network Authentication

A hash function h is referred to as homomorphic if for $x = \sum_i x_i$ it holds that $h(x) = \sum_i h(x_i)$. Homomorphic hash functions lend themselves naturally to the random (non-coherent) network coding scheme of

Ho *et al.* [26]. A node receiving information y_e, and coefficients $\{\alpha_i\}$ (that in the error free scenario should satisfy $y_e = \sum \alpha_i x_i$ for source information x_i) may check if $h(y_e) = \sum \alpha_i h(x_i)$ and so authenticate the received information. Here local information $h(x_i)$ is assumed to be known at internal nodes of the network. Indeed, if h is homomorphic and it is computationally *hard* for a given y to compute x such that $h(x) = y$, then a corrupted y_e will, with high probability, fail the authentication check.

Given the outline above it is natural to study the requirements from the local information $h(x_i)$ specified above. In the works of Krohn *et al.* [41] and Gkantsidis and Rodriguez [21], the hashes of the source information $h(x_i)$ are assumed to be reliably communicated to internal nodes of the network (otherwise, an adversary able to forge this information may indeed inject fake messages that will pass the internal node authentication process). Hence, a centralized trusted authority is assumed to provide these hashes. The security of the communication scheme suggested in [21, 41] is based on the hardness of the Discrete-Log problem.

In the works of Charles *et al.* [11], Zhao *et al.* [13], and Boneh *et al.* [8] the need for a reliable channel to distribute the hash values used for authentication is obviated using the notion of public key cryptography. In [11], the communication scheme suggested is based on the hardness of Discrete-Log and the computational co-Diffie-Hellman problem on elliptic curves. Zhao *et al.* [13] present a scheme based on *linear subspace authentication* which prevents the adversarial jammer to inject a fake message v into the network given that v is not in the space V spanned by the source information. Their scheme relies solely on the hardness of Discrete Log. Finally, in [8], two schemes based on the linear subspace authentication paradigm of [13] are presented. Boneh *et al.* [8] show that both schemes have public key sizes that are essentially optimal for this authentication paradigm. The first scheme of [8] is a homomorphic one and is based on the computational Diffie-Hellman assumption, while the second scheme is a non-homomorphic variation of the schemes of [41] and [13] which is based on the Discrete-Log problem.

Theorem 10 (In-Network Authentication). *Let G be an acyclic network with (error free) cut capacity C. Using in-network authentication, the capacity in the presence of a computationally bounded adversary that may jam up to z_O of the links of G is $C - z_O$.*

End-to-End Authentication

Similar to the works mentioned above, in [54] Nutman and Langberg also consider enhancing the (non–coherent) network coding scheme of Ho *et al.* [26]. However, in the communication scheme presented in [54] internal nodes of the network follow the exact same protocol as specified in [26]—and are thus oblivious to the presence of an adversarial jammer. The only changes made with respect to [26] are in the encoding and decoding procedures of the source and terminals.

To be more precise, the protocol of [54] builds on the non–coherent schemes of Jaggi *et al.* [27, 28] (which in turn builds on [26]) and has the following overall structure. In [27, 28], a non–coherent communication scheme of rate $C - z_O$ in the presence of an unconditional jammer (with unlimited computational power) that controls $C - z_O$ links is presented. The rate of $C - z_O$ is not possible in light of the discussion in Section 4.1 and can only be obtained under additional assumptions. Indeed in [27, 28], the rate $C - z_O$ is obtained under the additional assumption that the source and terminal nodes share a *low rate* side channel in which they may communicate a short secret (which is not known to the adversarial jammer). The analysis in [27, 28] is based on the observation that allowing *list decoding* (as opposed to *unique* decoding) at terminal nodes, rate $C - z_O$ is achievable in the presence of a jammer controlling z_O links (see Theorem 9). Once such a list is obtained, each terminal may pick the correct element from its list using the secret side information transmitted. The secrecy of the side information is crucial to avoid the jammer from imposing *tailor-made* errors that will imply certain lists at terminal nodes that cannot be disambiguated using the side information.

With the list decoding results of [27, 28] in mind (or any other list decodable scheme such as [47]), [54] considers the following natural modification. Instead of transmitting the side information of [27, 28] over a side channel (which is not present in the current model), [54] encrypts this information using any (not necessarily homomorphic) public key encryption scheme and transmits the encrypted side information over the network. Assuming the jammer cannot break the encryption scheme ensures that the side information remains secret, however the side information still needs to be transmitted to the terminals reliably. To attain this goal, [54] uses any one of the encoding schemes from Section 4.2 on the encrypted side information. Using the fact that the side information is of low rate, time sharing between the encoded side information and the coded source information yields rate $C - z_O$. We note that to ensure reliable communication of the side information, [54] requires that $C > 2z_O$. This last condition is proven in [54] to be necessary (under certain assumptions).

Theorem 11 (End–to–End Authentication). *Let G be an acyclic network with (error free) cut capacity C. Using end-to-end authentication, the capacity in the presence of a computationally bounded adversary that may jam up to z_O of the links of G is $C - z_O$ for $C > 2z_O$.*

5. SECRET TRANSMISSION IN THE PRESENCE OF EAVESDROPPING AND JAMMING ADVERSARIES

5.1. The Coherent Case

In this section we consider the interplay between eavesdropping and jamming. As we saw in Section 3, to protect a message against an eavesdropper that can listen to z_I links requires a rate-loss of at least z_I. We also saw that distributed low-complexity schemes with this rate-loss exist and achieve a secrecy rate of $C - z_I$. They do this essentially by linearly mixing a random message of rate z_I with the source message of rate $C - z_I$. Thus, these schemes can be thought of as a one-time pad ([64]) combined with network coding.

Next, in Section 4 we have seen that a network with a hidden adversarial jammer who observes all transmissions, and can jam z_O links, can effectively reduce the rate at which information can be transmitted from the source to the destination, down to $C - 2z_O$. Further, there are distributed low-complexity schemes that achieve this rate. These schemes can then be thought of as converting an error-prone operator channel of capacity C into an error-free operator channel of capacity $C - 2z_O$.

In scenarios where the adversary can only observe z_I transmissions in the network and jam z_O links, it is natural to ask what the best achievable rates of secret and reliable communication are. In the case with zero errors and single-letter coding, the work of [52] shows this to be the "natural" combination of the two above bounds, for an overall rate of $C - 2z_O - z_I$. They prove this by similar techniques as those used to bound the rates in the previous two sections. This bound was extended by [72] and [75] to zero-error block length coding as well.[7] Also, algorithms meeting these bounds are presented for the coherent case in [51] (for block coding) and [52] (for single-letter coding). These algorithms essentially work by merging the algorithms in the previous two sections—first they construct a coding scheme that converts the error-prone operator channel into an error-free operator channel of rate $C - 2z_O$, and on this channel they overlay a "one-time pad + network coding" scheme that ensures secrecy against a wiretapping adversary, which further reduces the rate to the overall rate of $C - 2z_O - z_I$. This leads to the following theorem.

Theorem 12 ([51, 52, 72, 75]). *The maximal rate at which secret information can be reliably communicated (with zero-error) over a network containing a hidden adversary who can eavesdrop on z_I links and jam z_O links is $C - 2z_O - z_I$.*

Interestingly, if one relaxes the requirement to zero-error to one of "small" error (asymptotically small in the field size or block length), then the upper bound of $C - 2z_O - z_I$ no longer holds—only a bound of $C - z_O - z_I$ can be shown. And in fact, as we shall see in the next subsection, this higher rate is in fact achievable with low-complexity code designs.

5.2. The Non-Coherent Case

The work in [72, 75] extends the results of Section 5.1 to give *universal* code designs. That is, given an arbitrary linear network code such that the rank of the linear transform is C, [72, 75] present an end-to-end scheme that treats the network code as an operator channel, and achieves the secrecy rate of $C - 2z_O - z_I$ as in Section 5.1. These constructions are based on rank-metric codes—it is shown that such codes are good not just for error correction as in Section 4.2, but also simultaneously for secrecy-preserving linear mappings at the source.

Further work in [86] demonstrated that as long as the sum of the adversary's jamming rate z_O and his eavesdropping rate z_I is less than the network

---

[7] In fact [72] proves the more general lower bound of $C - 2z_O - z_I - \rho$, where ρ is the number of (possibly adversarially located) erasures.

capacity C, (i.e., $z_O + z_I < C$), there exist codes with low computational complexity that can communicate (with vanishingly small error probability) a single bit correctly and without leaking any information to the adversary. This is then combined with a "secret-sharing" result of [27, 28] to design codes that allow communication at the optimal source rate of $C - z_O + z_I$ while keeping the communicated message secret from the adversary. In particular, the secret-sharing result of [27, 28] implies:

Theorem 13 ([27, 28]). *If in a network containing a hidden adversary who can jam at most z_O links, ϵn bits (for any fixed $\epsilon > 0$) can be secretly and reliably transmitted from the source to the destination, then in fact $(C - z_O - z_I)n$ bits can be secretly and reliably transmitted from the source to the destination.*

The main idea behind Theorem 13 is as follows. If the source node generates a "small" secret linear hash of its information and sends it to the receiver over a secret and reliable channel, then, using the linear list decoding result of Theorem 9, with high probability the receiver is able to refine the list down to a single element.

It only remains to describe a protocol to secretly and reliably share a bit over the network (one that may emulate a secret and reliable channel). To do this [86] uses the following straightforward "rank modulation" protocol. If the bit to be shared is a 0, then the source's message is a matrix (over a short block length, and hence asymptotically negligible in the true block length corresponding to the packet size) of rank $C - z_O - 1$; else its message is a random matrix of rank C. The decoder decodes by estimating the rank of the received matrix. If it equals C, it decodes the secret bit as 1, else it decodes to 0.

To check that the above protocol succeeds with high probability one needs to check both its secrecy and reliability. Secrecy is guaranteed since the adversary eavesdrops on at most z_I transmissions, which, due to the random linear mixing in the network and the constraint that $C - z_O - z_I > 0$, are not enough for it to be able to distinguish between a source message of rank $C - z_O - 1$, and a source message of rank C packets. Reliability is due to the following two arguments. First, since the adversary can inject at most z_O packets, if the source's message was 0 and so it transmits a matrix of rank $C - z_O - 1$, the rank of the received matrix must still be less than C. Conversely, if the source's message was 1 and hence it transmitted a truly random matrix of rank C, since the adversary does not know what this matrix is, the probability that it is able to reduce the rank of the received matrix is small.

6. SOME OTHER VARIANTS

We summarize here some of the other work on topics related to secure and reliable communication over networks, which do not fall neatly into previous sections.

- The work in references [36, 37, 63] considers network error-correction problems (with and without feedback) in scenarios where links have unequal capacities—a complete characterization of achievable rates in this case is still open.
- Kosut *et al.* [55, 56] consider the problem where *nodes* rather than edges are adversarially controlled. Here, again, the rate-region is yet to be fully characterized. Reliable communication using network coding in the presence of untrusted nodes is also considered in [82].
- Multiple-access variants of network error-correction have been considered in references [68–70, 81, 84].
- As an analog of the classical algorithms for point-to-point channels considered in [22], the work of [47] presents non-trivial list-decoding algorithms of network error correcting codes.
- The work of [59] considers the problems of reliability and secrecy for distributed data storage.
- The problem of finding the actual location of errors in the network has been considered in, among other works references [15, 16, 19, 23, 65–67, 85].
- In [45, 80], a **S**ecure **P**ractical **N**etwork **C**oding scheme (SPOC) is suggested that allows private communication against a computationally bounded adversary that may eavesdrop on all communication transmitted over the network. At its core, SPOC runs a modified variant of random linear coding [26] in which the *header* of each packet (containing the coding coefficients) is encrypted and unknown to the adversary, while the body of the packet (containing the encoded information via network coding) is sent in the clear.
- The authors of references [33–35] consider error detection in wireless networks in which adversarial nodes may behave maliciously. Using the

algebraic watchdog scheme, upstream nodes can detect malicious behaviors probabilistically by taking advantage of the broadcast nature of the wireless medium.

7. DISCUSSION

This chapter gives a brief summary of the coding schemes used in multicast network coding in the presence of passive and active jammers. We have seen that non-coherent secure communication at rate $C - z_I$ is possible in the presence of a passive eavesdropper that controls z_I links of the network. This rate is the best possible, even when considering coherent communication schemes. For active jammers, we have shown that non-coherent reliable communication at rate $C - 2z_I$ is possible in the presence of a jammer that controls z_O links of the network. If the jammer is computationally limited, a higher rate of $C - z_O$ is achievable. As before, these rates are the best possible, even when considering coherent communication schemes. Finally, when communicating in the presence of adversaries that may jam z_O links and eavesdrop on z_I links, communication which is both secure and reliable is possible at a tight rate of $C - 2z_O - z_I$ (or $C - z_O - z_I$ once one allows a small probability of error). The algorithmic techniques presented cover several paradigms and include tools from the study of combinatorics, linear algebra, and coding theory. The chapter at hand has addressed the task of multicast in acyclic networks. Understanding the power of network coding in a more general setting with or without adversaries remains an intriguing field of study that will surely evolve over the decades to come.

ACKNOWLEDGMENTS

The authors would like to thank Danilo Silva for many helpful discussions during the preparation of this chapter.

REFERENCES

[1] R. Ahlswede, N. Cai, S. R. Li, and R. W. Yeung. Network information flow. *IEEE Transactions on Information Theory*, 46(5): 1204–1216, July 2000.
[2] H. Balli, X. Yan, and Z. Zhang. Error correction capability of random network error correction codes. In *Proc. International Symposium on Information Theory*, Sept. 2007.

[3] H. Balli, X. Yan, and Z. Zhang. On randomized linear network codes and their error correction capabilities. *IEEE Transactions on Information Theory*, 55(7): 3148–3160, 2009.

[4] N. Baric and B. Pfitzmann. Collision-free accumulators and failstop signature schemes without trees. In *Advances in Cryptology EUROCRYPT*, 1997.

[5] M. Bellare and D. Micciancio. A new paradigm for collision-free hashing: Incrementality at reduced cost. In *Advances in Cryptology EUROCRYPT*, 1997.

[6] J. Benaloh and M. de Mare. One-way accumulators: A decentralized alternative to digital sinatures. In *Advances in Cryptology EUROCRYPT*, 1993.

[7] K. Bhattad and K. R. Narayanan. Weakly secure network coding. *In proceedings of NetCod*, 2005.

[8] D. Boneh, D. Freeman, J. Katz, and B. Waters. Signing a linear subspace: Signature schemes for network coding. In *12th International Conference on Practice and Theory in Public Key Cryptography*, pages 68–87, 2009.

[9] N. Cai and R. W. Yeung. Secure network coding. In *Proceedings of International Symposium on Information Theory*, Lausanne, Switzerland, June 2002.

[10] N. Cai and R. W. Yeung. Network error correction, part II: Lower bounds. *Commun. Inf. Syst*, 6(1): 37–54, 2006.

[11] D. Charles, K. Jain, and K. Lauter. Signatures for network coding. In *Proceedings of the Fortieth Annual Conference on Information Sciences and Systems*, Princeton, NJ, USA, 2006.

[12] D. Chaum, E. van Heijst, and B. Pfitzmann. Cryptographically strong undeniable signatures, unconditionally secure for the signer. In *Advances in Cryptology CRYPTO*, 1991.

[13] M. Médard, F. Zhao, T. Kalker, and K. J. Han. Signatures for content distribution with network coding. In *Proceedings of International Symposium on Information Theory (ISIT 2007)*, pages 556–560, 2007.

[14] J. Feldman, T. Malkin, C. Stein, and R. A. Servedio. On the capacity of secure network coding. In *Proceedings of 42nd Annual Allerton Conference on Communication, Control, and Computing*, Monticello, IL, 2004.

[15] C. Fragouli and A. Markopoulou. A network coding approach to network monitoring. In *Proc. of the 43nd Allerton Conference*, 2005.

[16] C. Fragouli, A. Markopoulou, and S. Diggavi. Topology inference using network coding. In *Proc. of the 44nd Allerton Conference*, 2006.

[17] E. M. Gabidulin. Theory of codes with maximum rank distance. *Probl. Inform. Transm.*, 21(1): 1–12, 1985.

[18] E. N. Gilbert. A comparison of signalling alphabets. *Bell Systems Technical Journal*, 31: 504–522, 1952.

[19] M. Gjoka, C. Fragouli, P. Sattari, and A. Markopoulou. Loss tomography in general topologies with network coding. In *Proc. of IEEE Globecom*, 2005.

[20] C. Gkantsidis and P. Rodriguez. Network coding for large scale content distribution. In *Proceedings of IEEE Conference on Computer Communications (INFOCOM)*, Miami, March 2005.

[21] C. Gkantsidis and P. Rodriguez. Cooperative security for network coding file distribution. In *Proceedings of IEEE Conference on Computer Communications (INFOCOM)*, pages 1–13, Barcelona, April 2006.

[22] V. Guruswami and M. Sudan. Improved decoding of Reed-Solomon and algebraic-geometric codes. *IEEE Transactions on Information Theory*, 45: 1757–1767, September 1999.

[23] L. Hailiang, H. Guangmin, Q. Feng, and Y. Zhihao. Network topology infer-
      ence based on traceroute and tomography. In *Proc. of International Conference on
      Communications and Mobile Computing*, 2009.
[24] T. Ho., R. Kötter, M. Médard, D. Karger, and M. Effros. The benefits of coding over
      routing in a randomized setting. In *IEEE International Symposium on Information Theory
      (ISIT)*, page 442, Yokohama, July 2003.
[25] T. Ho, B. Leong, R. Koetter, M. Médard, M. Effros, and D. R. Karger. Byzan-
      tine modification detection in multicast networks using randomized network coding.
      IEEE Transactions on Information Theory, 54(6): 2798–2803, 2008.
[26] T. Ho, M. Médard, R. Kötter, D. R. Karger, M. Effros, J. Shi, and B. Leong. A random
      linear network coding approach to multicast. *IEEE Transactions on Information Theory*,
      52(10): 4413–4430, 2006.
[27] S. Jaggi, M. Langberg, S. Katti, T. Ho, D. Katabi, and M. Médard. Resilient network
      coding in the presence of Byzantine adversaries. In *Proc. 26th IEEE Int. Conf. on
      Computer Commun.*, pages 616–624, Anchorage, AK, May 2007.
[28] S. Jaggi, M. Langberg, S. Katti, T. Ho, D. Katabi, M. Médard, and M. Effros.
      Resilient network coding in the presence of Byzantine adversaries. *IEEE Transactions
      on Information Theory*, 54(6): 2596–2603, June 2008.
[29] S. Jaggi, P. Sanders, P. A. Chou, M. Effros, S. Egner, K. Jain, and L. Tolhuizen. Poly-
      nomial time algorithms for multicast network code construction. *IEEE Transactions on
      Information Theory*, 51(6): 1973–1982, June 2005.
[30] K. Jain. Security based on network topology against the wiretapping attack. *IEEE
      Wireless Communications*, pages 68–71, Feb 2004.
[31] R. Johnson, D. Molnar, D. Song, and D. Wagner. Homomorphic signature schemes.
      In *Progress in Cryptology CT RSA*, 2002.
[32] S. Katti, H. Rahul, D. Katabi, W. Hu M. Médard, and J. Crowcroft. XORs in the
      Air: Practical Wireless Network Coding. In *ACM SIGCOMM*, Pisa, Italy, 2006.
[33] M. Kim, M. Médard, and J. Barros. A multi-hop multi-source algebraic watchdog. In
      IEEE Information Theory Workshop (ITW), Dublin, Ireland, August 2010.
[34] M. Kim, M. Médard, and J. Barros. Algebraic watchdog: Mitigating misbehavior
      in wireless network coding. Submitted to *IEEE Journal on Selected Areas in
      Communications (JSAC)*, Advances in Military Networking and Communications,
      http://arxiv.org/abs/1011.3879, November 2010.
[35] M. Kim, M. Médard, J. Barros, and R. Kötter. An algebraic watchdog for wireless
      network coding. In *IEEE International Symposium on Information Theory (ISIT)*, June
      2009.
[36] S. Kim, T. Ho, M. Effros, and S. Avestimehr. Network error correction with unequal
      link capacities: Capacities and upper bound. *In preparation for submission to the IEEE
      Transactions on Information Theory*, 2009.
[37] S. Kim, T. Ho, M. Effros, and S. Avestimehr. New results on network error correc-
      tion: Capacities and upper bound. In *Proceedings of Information Theory and Applications
      Workshop, UCSD*, San Diego, CA, 2010.
[38] R. Kötter and F. R. Kschischang. Coding for errors and erasures in random network
      coding. In *Proc. IEEE Int. Symp. Information Theory*, pages 791–795, Nice, France,
      June 24–29, 2007.
[39] R. Kötter and F. R. Kschischang. Coding for errors and erasures in random network
      coding. *IEEE Transactions on Information Theory*, 54(8): 3579–3591, Aug. 2008.
[40] R. Kötter and M. Médard. An algebraic approach to network coding. *IEEE/ACM
      Transactions on Networking*, 11(5): 782–795, October 2003.

[41] Maxwell N. Krohn, Michael J. Freedman, and David Mazires. On-the-fly verification of rateless erasure codes for efficient content distribution. In *Proceedings of the IEEE Symposium on Security and Privacy*, pages 226–240, Oakland, California, 2004.

[42] M. Langberg, A. Sprintson, and J. Bruck. The encoding complexity of network coding. *IEEE Transactions on Information Theory (a joint special issue with IEEE/ACM Transactions on Networking on Networking and Information Theory)*, 52(6): 2386–2397, 2006.

[43] M. Langberg, A. Sprintson, and J. Bruck. Network coding: A computational perspective. *IEEE Transactions on Information Theory*, 55(1): 147–157, 2009.

[44] S.-Y. R. Li, R. W. Yeung, and N. Cai. Linear network coding. *IEEE Transactions on Information Theory*, 49(2): 371–381, 2003.

[45] L. Lima, J. P. Vilela, J. Barros, and M. Médard. An information-theoretic cryptanalysis of network coding - is protecting the code enough? In *Proceeding of the International Symposium on Information Theory and its Applications (ISITA)*, 2008.

[46] D. S. Lun, M. Médard, and R. Kötter. Efficient operation of wireless packet networks using network coding. In *International Workshop on Convergent Technologies (IWCT)*, Oulu, Finland, 2005.

[47] H. Mahdavifar and A. Vardy. Algebraic list-decoding on the operator channel. In *Proc. of International Symposium on Information Theory (ISIT)*, Austin, TX, USA, June 13–19 2010.

[48] F. J. McWilliams and N. J. A. Sloane. *The Theory of Error-Correcting Codes*. North-Holland, 1977.

[49] S. Micali and R. Rivest. Transitive signature schemes. In *Progress in Cryptology CT RSA*, 2002.

[50] A. Montanari and R. Urbanke. Coding for network coding. Available online at http://arxiv.org/abs/0711.3935.

[51] C. K. Ngai and S. Yang. Deterministic secure error-correcting (SEC) network codes. In *Proc. IEEE. Information Theory Workshop*, pages 96–101, Bergen, Norway, June 24–29, 2007.

[52] C.-K. Ngai and R. W. Yeung. Secure error-correcting (SEC) network codes. In *Workshop on Network Coding, Theory and Applications (NetCod)*, pages 791–795, Lausanne, Switzerland, June 2009.

[53] C. K. Ngai, R. W. Yeung, and Z. Zhang. Network Generalized Hamming Weight. *In Proc. 2009 Workshop on Network Coding, Theory, and Applications*, 2009.

[54] L. Nutman and M. Langberg. Adversarial models and resilient schemes for network coding. In *Proceedings of International Symposium on Information Theory*, pages 171–175, 2008.

[55] L. Tong, O. Kosut, and D. Tse. Nonlinear network coding is necessary to combat general byzantine attacks. In *Proc. of 47th Annual Allerton Conference on Communication, Control, and Computing*, October 2009.

[56] L. Tong, O. Kosut, and D. Tse. Polytope codes against adversaries in networks. In *Proc. of International Symposium on Information Theory (ISIT)*, Austin, TX, USA, June 13–19, 2010.

[57] L. H. Ozarow and A. D. Wyner. The wire-tap channel II. *Bell Syst. Tech. Journ.*, 63: 2135–2157, 1984.

[58] L. H. Ozarow and A. D. Wyner. Wire-tap channel II. *In Proceedings of the EURO-CRYPT 84 workshop on advances in cryptology: Theory and application of cryptographic techniques*, pages 33–51, 1985.

[59] S. Pawar, S. El-Rouayheb, and K. Ramchandran. On secure distributed data storage under repair dynamics. In *Proc. IEEE Int. Symp. Information Theory*, June 13–19 2010.

[60] T. P. Pedersen. Non-interactive and information-theoretic secure verifiable secret sharing. In *Advances in Cryptology CRYPTO*, 1991.

[61] S. E. Rouayheb, E. Soljanin, and A. Sprintson. Secure network coding for wiretap networks of type II. *Manuscript available on arXiv,org*, 2009.

[62] S. E. Rouayheb and E. Y. Soljanin. On Wiretap Networks II. *In proceedings of IEEE International Symposium on Information Theory*, pages 551–555, 2007.

[63] M. Effros, S. Kim, T. Ho, and S. Avestimehr. Network error correction with unequal link capacities. In *Proceedings of 47th Annual Allerton Conference on Communication, Control, and Computing*, Monticello, IL, 2008.

[64] C. E. Shannon. Communication theory of secrecy systems. *Bell System Technical Journal*, 28(4): 656–715, 1949.

[65] J. M. Siavoshani, C. Fragouli, and S. Diggavi. Subspace properties of randomized network coding. In *Proc. of IEEE ITW*, 2007.

[66] M. Jafari Siavoshani, C. Fragouli, and S. Diggavi. On locating byzantine attackers. In *Network Coding Workshop: Theory and Applications*, 2008.

[67] M. Jafari Siavoshani, C. Fragouli, S. Diggavi, and C. Gkantsidis. Bottleneck discovery and overlay management in network coded peer-to-peer system. In *Proc. of SIGCOMM Workshop on Internet Network Management*, 2007.

[68] M. J. Siavoshani, C. Fragouli, and S. Diggavi. Noncoherent multisource network coding. In *Proc. IEEE Int. Symp. Information Theory*, pages 817–821, Toronto, Canada, July 6–11, 2008.

[69] M. J. Siavoshani, C. Fragouli, and S. Diggavi. Code construction for multiple sources network coding. In *Proc. of the MobiHoc*, 2009.

[70] M. J. Siavoshani, S. Mohajer, C. Fragouli, and S. Diggavi. On the capacity of noncoherent network coding. In *Proc. of the IEEE International Symposium on Information Theory*, 2009.

[71] D. Silva and F. R. Kschischang. Using rank-metric codes for error correction in random network coding. In *Proc. IEEE Int. Symp. Information Theory*, pages 796–800, Nice, France, June 24–29, 2007.

[72] D. Silva and F. R. Kschischang. Universal secure network coding via rank-metric codes. *IEEE Transactions on Information Theory*, 57(2): 1124–1135, 2011.

[73] D. Silva and F. R. Kschischang. Fast encoding and decoding of Gabidulin codes. In *Proc. IEEE Int. Symp. Information Theory*, Seoul, Korea, June 2009.

[74] D. Silva and F. R. Kschischang. On metrics for error correction in network coding. *IEEE Transactions on Information Theory*, 55(12): 5479–5490, 2009.

[75] D. Silva and F. R. Kschischang. Universal secure error control schemes for network coding. In *Proc. IEEE Int. Symp. Information Theory*, Austin, TX, USA, June 13–19, 2010.

[76] D. Silva, F. R. Kschischang, and R. Kötter. Capacity of random network coding under a probabilistic error model. In *Proc. 24th Biennial Symposium on Communications*, Kingston, ON, Canada, June 2008.

[77] D. Silva, F. R. Kschischang, and R. Kötter. A rank-metric approach to error control in random network coding. *IEEE Transactions on Information Theory*, 54(9): 3951–3967, 2008.

[78] Danilo Silva and Frank R. Kschischang. Universal weakly secure network coding. In *Proc. Inform. Theory Workshop on Networking and Inform. Theory*, pages 281–285, Volos, Greece, June 10-12, 2009.

[79] R. R. Varshamov. Estimate of the number of signals in error correcting codes. *Dokl. Acad. Nauk*, 117: 739–741, 1957.

[80] J. P. Vilela, L. Lima, and J. Barros. Lightweight security for network coding. In *Proc. of the IEEE International Conference on Communications (ICC 2008)*, Beijing, China, May 2008.

[81] S. Vyetrenko, T. Ho, M. Effros, J. Kliewer, and E. Erez. Rate regions for coherent and noncoherent multisource network error correction. In *Proc. of International Symposium on Information Theory (ISIT)*, Seoul, Korea, June 2009.

[82] D. Wang, D. Silva, and F. R. Kschischang. Robust network coding in the presence of untrusted nodes. *IEEE Transactions on Information Theory*, 56(9): 4532–4538, 2010.

[83] V. K. Wei. Generalized Hamming Weight for linear codes. *IEEE Trans. Inform. Theory*, 37(5): 1412–1418, 1991.

[84] H. Yao, T. K. Dikaliotis, S. Jaggi, and T. Ho. Multiple access network information-flow and correction codes. In *Proc. IEEE Information Theory Workshop*, Dublin, Ireland, 2010.

[85] H. Yao, S. Jaggi, and M. Chen. Passive network tomography for erroneous networks: A network coding approach. Under submission to the IEEE Transactions on Information Theory, 2010.

[86] H. Yao, D. Silva, S. Jaggi, and M. Langberg. Network codes resilient to jamming and eavesdropping. In *Proc. Workshop on Network Coding Theory and Applications*, Toronto, Canada, June 9–11 2010.

[87] R. W. Yeung and N. Cai. Network error correction, part I: Basic concepts and upper bounds. *Commun. Inf. Syst*, 6(1): 19–36, 2006.

[88] Z. Zhang. Linear network error correction codes in packet networks. *IEEE Transactions on Information Theory*, 54(1): 209–218, January 2008.

CHAPTER 8

Network Coding and Data Compression

Mayank Bakshi[1], Michelle Effros[1], Tracey Ho[1] and **Muriel Médard[2]**

[1]Department of Electrical Engineering, California Institute of Technology, Pasadena, CA, USA;
[2]Department of Electrical Engineering, Massachusetts Institute of Technology, Cambridge, MA, USA

Contents

Abstract

In this chapter the problem of communicating statistically dependent sources over networks is considered, focusing in particular on the case of lossless reconstruction over acyclic networks of noise-free links. In contrast to a purely network coding setup with independent sources, optimal code design for this setup involves both network coding as well as data compression aspects. Calculating the set of achievable rates for this problem is a difficult task in general, though some continuity properties of the rate region are known, and a general outer bound can be obtained from cut-set arguments. Even though the cut-set bounds are not achievable for all networks, they are achievable for multicast networks, i.e., networks where all sources are demanded at all sinks. The cut-set bounds are also achievable for an extension of the multicast network, where, in addition to the desired sources, the network may have side information present at some of the sinks. For both these cases, cut-set bounds are achievable via linear network codes. Finally, some practical approaches to combining network coding and compression are discussed.

Network Coding. DOI: 10.1016/B978-0-12-380918-6.00008-1
Copyright © 2012 Elsevier Inc. All rights reserved.

Keywords: Distributed compression, joint source-network coding, lossless, multicast, network source coding, side information.

1. INTRODUCTION

The problems of network source coding and network coding are closely related. In each, we are given a network of noiseless, capacitated links. Each node may observe one or more sources and request one or more communication demands. At the highest level, the question is whether we can design codes for describing the sources over the network in a manner that allows all nodes to meet their corresponding demands.

In network source coding, the sources originating at different nodes of the network are typically assumed to be drawn from an arbitrary (typically known) distribution, i.e. arbitrary statistical dependencies are allowed. The source alphabets may be discrete or continuous. Demands may be either sources available at other nodes in the network (e.g., [1, 2]) or functions of those sources (e.g., [3, 4]). Associated with each demand is a required level of reconstruction fidelity. Fidelity requirements range from zero-error reconstruction (where the probability that the source differs from its reconstruction is precisely 0—see, for example, [5, 6]), to lossless reconstruction (where that same probability can be made arbitrarily small but may not be precisely 0—see, for example, [1]), to lossy reconstruction (where a distortion measure is specified, and a constraint is placed on the maximal expected distortion—see, for example, [7, 8]). The majority of the network source coding literature treats special cases of the network source coding problem—fixing a network topology and collection of demands and then characterizing—in whole or in part—the space of link capacity values for which it is possible to meet the given demands. (See, for example, [1, 2, 7–11].)

In the work that initiated the field of network coding [12], network coding is characterized as a data transmission problem rather than a data representation problem. As a result, in the network coding literature, the sources traversing the network are most commonly treated as independent and uniformly distributed random variables on finite alphabets. Both lossless (e.g., [12]) and zero-error (e.g., [13, 14]) demands are treated in the literature. While the majority of the network coding literature treats independent and uniform sources, generalizations to the case of dependent sources are also considered (e.g., [15]). In this case, network coding and lossless network source coding are equivalent. The majority of the network coding literature treats general coding scenarios rather than specific topologies. The multicast and multi-source multicast problems, where one or more transmitters send the same information to all receivers in a fixed collection of receivers, are examples of such coding scenarios (see, for example, [12, 15]). In these kinds of coding scenarios, results are derived for networks with arbitrary topologies but fixed types of demands.

The problems of network coding and network source coding are not only conceptually similar, they are, in some sense, inseparable. References [16] and [17] give non-multicast and multicast examples, respectively, of communication scenarios where network source coding and network coding must be done jointly in order to achieve the optimal end-to-end communication performance. In these scenarios, separation would imply that each source could be compressed once at its originating node and then transmitted through the network without knowledge of its statistics by intermediate nodes or recompression elsewhere in the network. Figure 8.1(a) reproduces an instance of the example from [16]. Here sources X_1 and X_2 are statistically dependent random variables with entropies $H(X_1) = H(X_2) = 3$ and joint entropy $H(X_1, X_2) = 4 < H(X_1) + H(X_2)$. Each edge has capacity $1 + \epsilon$ for some $\epsilon > 0$. In order to know how to properly compress sources X_1 and X_2 in this example, nodes 1 and 2 would have to know something about the network topology. Equivalently, the outputs of nodes 1 and 2 must remain statistically dependent, and in order to determine how to recode those values, nodes 3 and 4 would have to know something about the joint statistics of the sources. Figure 8.1(b) reproduces the two-transmitter, three-receiver multicast example from [17]. Here sources X_1 and X_2 are statistically dependent random variables with entropies $H(X_1) = H(X_2) = 2$ and joint entropy $H(X_1, X_2) = 3$. Again, each edge has capacity $1 + \epsilon$. Intuitively, the butterfly transmits one bit from each source to the receivers at nodes 11 and 12. Those receivers each get a third bit from node 5 or 8, so nodes 5 and 8 cannot carry the same information that traverses the butterfly. Unfortunately, in a separated strategy, this leaves node 3 unable to decode since it cannot separate the bits that are tangled together for transmission through the

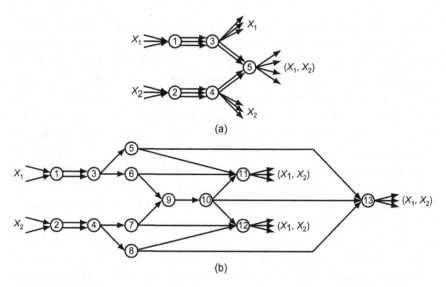

Figure 8.1 Example networks where the separation between source and network coding fails. (a) An instance of the non-multicast example from [16]. In this example, sources X_1 and X_2 are jointly distributed with individual entropies $H(X_1) = H(X_2) = 3$ and joint entropy $H(X_1, X_2) = 4$. (b) The two-transmitter, three-receiver multicast example from [17]. In this example, sources X_1 and X_2 are jointly distributed with individual entropies $H(X_1) = H(X_2) = 2$ and joint entropy $H(X_1, X_2) = 3$. In both examples, each edge has capacity $1 + \epsilon$ for some $\epsilon > 0$.

bottleneck of the butterfly. In contrast, a joint source–network code can reliably transmit each source to all intended receivers, as originally proved in [15] and discussed in Section 4.

Since even some very small network source coding problems remain incompletely understood [18] and the field of network coding is rife with examples of NP-hard problems [19–21], it is perhaps not surprising that our knowledge to date of the field of joint source–network coding remains limited. The remainder of this chapter restricts attention to lossless joint source–network problems for dependent sources on general network topologies. Section 2 gives the formal problem statement and defines the necessary notation. Section 3 gives a general overview on network capacity regions. Section 4 introduces both coding strategies and achievable rate regions for a few special cases of the general joint source–network coding problem where our knowledge to date is relatively rich.

2. MODEL AND NOTATION

We consider a directed acyclic network with noise-free links, represented as a graph $(\mathcal{V}, \mathcal{E})$ where $\mathcal{V}$ is the vertex set and $\mathcal{E} \subseteq (\mathcal{V} \times \mathcal{V})$ is the edge set. Let the capacity of an edge $e \in \mathcal{E}$ be denoted c_e. The set of incoming links (w, v) of a node v is denoted $\mathcal{I}(v)$ and the set of outgoing links (v, w) of v is denoted $\mathcal{O}(v)$.

There is a set of source nodes $\mathcal{S} \subset \mathcal{V}$. For simplicity and without loss of generality, we assume that each source node has no input edges.[1] Each source node $s \in \mathcal{S}$ observes a single random source process X_s taking values from a discrete alphabet $\mathcal{X}_s$; for simplicity we assume that X_s has an integer entropy rate of r_s bits per unit time with $\mathcal{X}_s = \mathbb{F}_2^{r_s}$. The vector of symbols $(X_s(i) : s \in \mathcal{S})$ at each time step i is drawn i.i.d. from a given joint distribution Q. Each node v demands a subset $\mathcal{D}(v) \subset \mathcal{S}$ of the sources in the network. Thus node v wishes to losslessly reconstruct $X_{\mathcal{D}(v)} = (X_s : s \in \mathcal{D}(v))$.[2]

A joint source-network code defines a mapping:

$$f_e^{(n)} : \mathcal{X}_s^n \to \{1, \ldots, 2^{n c_e}\}, \ e \in \mathcal{O}(s)$$

from source symbol X_s^n to output symbol U_e^n transmitted on each outgoing edge e of a source node $s \in \mathcal{S}$, a mapping:

$$f_e^{(n)} : \prod_{d \in \mathcal{I}(v)} \{1, \ldots, 2^{n c_d}\} \to \{1, \ldots, 2^{n c_e}\}, \ e \in \mathcal{O}(v)$$

from received symbols to output symbol U_e^n on each outgoing edge e of a non-source node v, and a decoder mapping:

$$g_t^{(n)} : \prod_{d \in \mathcal{I}(t)} \{1, \ldots, 2^{n c_d}\} \to \prod_{s \in \mathcal{D}(t)} \mathcal{X}_s^n$$

---

[1] If source node v has input edges, we can obtain a network with the same capacity region by creating a virtual source node v' with no input edges and an output edge (v', v) of infinite capacity. Making v' the source instead of v gives a network with the same capacity region that satisfies the constraint.

[2] Again for simplicity and without loss of generality, each source node observes a single source that is either desired in whole or not desired at all by each other node in the network. We handle multiple sources at a single source node by creating virtual nodes for each distinct source as described above.

from received symbols to decoded symbols for each sink node $t \in \mathcal{V}$ with $\mathcal{D}(t) \neq \emptyset$. Since the network is acyclic, the node mappings can be applied in topological order such that each node receives input symbols from all its incoming edges before applying the mappings corresponding to its outgoing edges.

We focus on lossless reconstruction, which requires that the probability of decoding error goes to zero as the coding block length n grows without bound. The goal of the source and network coding literatures is to characterize when a particular collection of sources can be losslessly transmitted across a particular network of noiseless links, and to design efficient communication strategies that achieve these limits in practice. While the underlying goal of determining demand feasibility is the same and both fields study this feasibility question by characterizing achievable rate regions, the definitions of "rate regions" employed in the two literatures differ. In both, the network topology is fixed. The source coding rate region typically fixes the source distribution and then characterizes the vector of link capacities ($c_e : e \in \mathcal{E}$) required to simultaneously satisfy all demands $\mathcal{D}(t)$, $t \in \mathcal{V}$. In contrast, the network coding literature typically fixes the link capacities and then describes the vector of rates ($r_s : s \in \mathcal{S}$) for which all demands $\mathcal{D}(t)$, $t \in \mathcal{V}$, can be simultaneously satisfied. We here employ the source coding characterization, formally defined below. This choice is motivated by the observation that for dependent sources, even the full vector of entropies ($H(X_{\mathcal{S}'}) : \mathcal{S}' \subseteq \mathcal{S}$) is insufficient to characterize the link capacities required to transmit the sources ($X_s : s \in \mathcal{S}$); this contrasts with the case for independent sources, where the source entropies r_s suffice for characterizing achievability of a given collection of demands [22, Corollary 2.3.3]. To illustrate this observation, consider the example shown in Fig. 8.2. Sources $X_{\mathcal{S}} = (X_1, X_2)$ and $\hat{X}_{\mathcal{S}} = (\hat{X}_1, \hat{X}_2)$ have identical entropy vectors:

$$(H(X_1), H(X_2), H(X_1, X_2)) = (H(\hat{X}_1), H(\hat{X}_2), H(\hat{X}_1, \hat{X}_2)).$$

When the inputs at nodes 1 and 2 are X_1 and X_2, respectively, then the demands $\mathcal{D}(3) = \{1\}$ can be satisfied at node 3. In contrast, when the inputs at nodes 1 and 2 are $\hat{X}_1$ and $\hat{X}_2$, respectively, then the demands $\mathcal{D}(3) = \{1\}$ cannot be satisfied at node 3, [2].

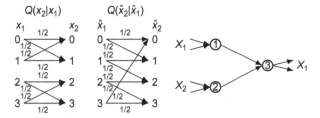

Figure 8.2 The rate region for sources (X_1, X_2) and $(\hat{X}_1, \hat{X}_2)$ may be different even when the entropy vectors $(H(X_1), H(X_2), H(X_1, X_2))$ and $(H(\hat{X}_1), H(\hat{X}_2), H(\hat{X}_1, \hat{X}_2))$ are identical, as shown by the given example. Sources X_1 and $\hat{X}_1$ are uniformly distributed on alphabet $\{0, 1, 2, 3\}$. Sources X_2 and $\hat{X}_2$ are drawn according to distributions $Q(x_2|x_1)$ and $Q(\hat{x}_2|\hat{x}_1)$, respectively, as defined above. The rate regions for the coded side information problem differ for these two sources. For example, when $p = 0.5$ and each link has capacity $1 + \epsilon$, the given demands can be met for source (X_1, X_2) but not for source $(\hat{X}_1, \hat{X}_2)$.

For a given network $(\mathcal{V}, \mathcal{E})$ with sources $\mathcal{S} \subseteq \mathcal{V}$ drawn from distribution Q and demands $\mathcal{D}(t) \subseteq \mathcal{S}$ for all $t \in \mathcal{V}$, we use $\mathcal{R}$ to describe the set of edge capacity vectors $\mathbf{c} = (c_e : e \in \mathcal{E})$ for which the given demands are achievable. We use $\mathcal{R}_L$ to denote the corresponding set of capacity vectors if we restrict $f_e^{(n)}$ for each $e \in \mathcal{E}$ to linear functions. The rate regions $\mathcal{R}$ and $\mathcal{R}_L$ are not fully characterized for all sources and all demands. Section 3 summarizes a variety of known properties and bounds for joint source-network source codes.

3. RATE REGION PROPERTIES FOR GENERAL JOINT SOURCE-NETWORK CODING

While a variety of special cases are well understood, very little is known about the achievable rate region for the general joint source-network coding problem. We here give a very brief overview of existing results.

Consider a network $(\mathcal{V}, \mathcal{E})$ with sources $\mathcal{S}$ and demands $\mathcal{D}(t)$ for all $t \in \mathcal{V}$. Suppose that for every source $s \in \mathcal{S}$ and every sink $t \in \mathcal{V}$ for which $s \in \mathcal{D}(t)$ there exists a path from node s to node t. Under mild conditions on the distortion measure, the lossless rate region $\mathcal{R}$ equals the lossy rate region evaluated at distortion 0 [23][22, Theorem 2.5.13]. Further, the rate region for $\mathcal{R}$ changes continuously in the underlying source distribution Q. This property is important since true source distributions are rarely, if ever, known in practice. Instead, we rely on empirical estimates of the source distribution based on past source samples. The continuity of the rate region in the source distribution implies that an asymptotically small error in the source distribution estimate yields an asymptotically small error in the corresponding rate region.

While rate regions are not solved in general and may be hard to calculate [19, 20], a variety of tools are available for bounding these rate regions under the assumption of statistically dependent sources. Chief among these are the cut-set bounds, described next. A cut between $\mathcal{S}'$ and t is a partition of $\mathcal{V}$ into two sets $\mathcal{A}$ and $\mathcal{V}\backslash\mathcal{A}$ such that $\mathcal{S}' \subset \mathcal{A}$ and $t \in \mathcal{V}\backslash\mathcal{A}$. The capacity of the cut is the sum of the capacities of the edges whose start node is in $\mathcal{A}$ and whose end node is in $\mathcal{V}\backslash\mathcal{A}$. For a given vector **c** of edge capacities, we denote by $m_{\mathcal{S}',t}(\mathbf{c})$ the minimum cut capacity between a subset of sources $\mathcal{S}' \subseteq \mathcal{S}$ and a sink t. For each cut $\mathcal{A}$ between source set $\mathcal{S}' \subseteq \mathcal{D}(t)$ and sink t, the minimal rate required for describing the sources in $\mathcal{S}'$ to sink t equals $H(X_{\mathcal{S}'}|X_{\mathcal{S}\backslash\mathcal{S}'})$—which is the rate required by the Slepian-Wolf theorem [24] for describing $X_{\mathcal{S}'}$ to a node that knows $X_{\mathcal{S}\backslash\mathcal{S}'}$ [25, Section 15.10]. Let $\mathcal{R}_C$ denote the set of edge capacity vectors **c** satisfying the cut-set bounds:

$$ m_{\mathcal{S}',t}(\mathbf{c}) > H(X_{\mathcal{S}'}|X_{\mathcal{S}\backslash\mathcal{S}'}) \ \forall \ \mathcal{S}' \subset \mathcal{D}(t), \ t \in \mathcal{V} \tag{1} $$

for a given network $(\mathcal{V}, \mathcal{E})$ and source distribution Q. The cut-set bounds provide an outer bound on the rate region for the lossless source coding problem. Precisely, for any fixed network $(\mathcal{V}, \mathcal{E})$ with source nodes $\mathcal{S}$ and demands $\mathcal{D}(t)$, and for any fixed source distribution Q:

$$ \mathcal{R}_L \subseteq \mathcal{R} \subseteq \mathcal{R}_C. $$

The cut-set bounds are not tight in general. For example, the coded side information problem illustrated by the network in Fig. 8.2 is an example of a very small network where the cut-set bounds cannot be achieved with equality for all possible distributions on sources (X_1, X_2). Section 4 describes a family of joint source-network coding problems in which cut-set bounds are tight. The discussion also treats a variety of mechanisms for approximating this optimal limiting performance in practice.

4. CAPACITY RESULTS FOR LOSSLESS MULTICAST

In this section we describe two settings in which the capacity region for joint source-network coding of statistically dependent sources is known. Both of these examples fall under the heading of multi-source multicast problems, where there exists some collection $\mathcal{A} \subseteq \mathcal{S}$ of sources and some collection $\mathcal{T}$ of sinks such that $\mathcal{D}(t) = \mathcal{A}$ for all $t \in \mathcal{T}$, and $\mathcal{D}(v) = \emptyset$ for all $v \notin \mathcal{T}$. In the first case, $\mathcal{A} = \mathcal{S}$, while in the second case $\mathcal{A}$ is a strict subset of $\mathcal{S}$ and the remaining sources serve as side information at the receivers.

4.1. No Side Information Scenario

We first consider the scenario with no side information, i.e., all the information in the network originates at the source nodes and is demanded by all the sinks. In this case, the capacity region is given by a cut-set characterization, and is achievable with linear network coding and nonlinear decoding.

This result is stated formally as follows.

Theorem 1. $\mathcal{R} = \mathcal{R}_L = \mathcal{R}_C$.

The region $\mathcal{R}_C$ defined by (1) corresponds to the Slepian-Wolf capacity region for each sink individually. The above theorem states that with linear network coding, any rate vector achievable for each sink individually is achievable for all sinks simultaneously. Note that different sinks can operate at different points in the Slepian-Wolf region, so statistically dependent sources can act as partial "mirror sites" for each other.

To show achievability of the region $\mathcal{R}_C$ with linear coding, we consider random vector linear coding over blocks consisting of nr_s bits from each source s. Specifically, the symbol on an outgoing edge e of a source node s is given by:

$$U_e^n = \mathbf{F}_{s,e} X_s^n, \ e \in \mathcal{O}(s)$$

and the symbol on an outgoing edge e of a non-source node v is given by:

$$U_e^n = \sum_{d \in \mathcal{I}(v)} \mathbf{F}_{d,e} U_d^n, \ e \in \mathcal{O}(v),$$

where $\mathbf{F}_{s,e}$ and $\mathbf{F}_{d,e}$ are binary matrices of dimensions $nr_s \times nc_e$ and $nc_d \times nc_e$ respectively, whose entries are chosen independently and uniformly at random from $\mathbb{F}_2$. The decoding function is nonlinear. Each sink maps its received symbols to a vector of decoded symbols that has minimum entropy or maximum Q-probability among all possible source symbol vectors consistent with the received symbols. The probability of decoding error at each sink can be upper bounded in terms of the cut capacities $m_{\mathcal{S}',t}(\mathbf{c})$ and the maximum source-receiver path length L, with the bound going to zero asymptotically with the block length n. We omit the proof for brevity, but it can be found in the appendix.

4.2. Side Information at Sinks

A generalization of the above scenario for which capacity is also known allows for side information at the sink nodes. The model is the same as that described in Section 2, except that we assume here that each sink node $t \in \mathcal{T}$ has no outgoing edges and observes a side information random process Y_t taking values from a discrete alphabet Y_t. The assumption that each sink node has no outgoing edges avoids the need to consider encoding and transmitting side information, which would be much more complicated. The vector of symbols $(X_s(i) : s \in \mathcal{S}, Y_t(i) : t \in \mathcal{T})$ at each time step i is drawn i.i.d. from a given joint distribution Q. Each sink demands all the sources' information but not the side information. The definition of $\mathcal{R}_C$ is modified accordingly to be the set of rate vectors satisfying the cut-set bounds:

$$m_{\mathcal{S}',t}(\mathbf{c}) > H(X_{\mathcal{S}'}|X_{\mathcal{S}\backslash\mathcal{S}'}, Y_t) \ \forall \ \mathcal{S}' \subset \mathcal{S}, t \in \mathcal{T}.$$

Theorem 2. $\mathcal{R} = \mathcal{R}_L = \mathcal{R}_C$.

Proof. Since every achievable rate vector $\mathbf{c}$ satisfies the cut-set bounds, it follows that $\mathcal{R} \subseteq \mathcal{R}_C$, and therefore $\mathcal{R}_L \subseteq \mathcal{R} \subseteq \mathcal{R}_C$.

To prove the equality, it suffices to show that $\mathcal{R}_L = \mathcal{R}_C$. Towards this, we show that a sufficient condition for a sequence of random linear codes to achieve vanishing error probability at each sink is $\mathbf{c} \in \mathcal{R}_C$.

For $t \in \mathcal{T}$, let $(\mathcal{V}_t, \mathcal{E}_t)$ be the network obtained by deleting all sinks except the sink t. Thus, $\mathcal{V}_t = (\mathcal{V}\backslash\mathcal{T}) \cup \{t\}$ and $\mathcal{E}_t = \mathcal{E}\backslash \cup t' \in J\backslash\{t\}I(t')$. Note that since Y_t is already known to the sink t, there is is no loss of optimality by treating the network $(\mathcal{V}_t, \mathcal{E}_t)$ as a multicast network with demands $(X_{\mathcal{S}}, Y_t)$.

For a rate vector $\mathbf{c}$, let $\mathbf{c}_t$ denote the restriction of $\mathbf{c}$ to $(\mathcal{V}_t, \mathcal{E}_t)$. Since there is no directed path from a node in $\mathcal{T} \setminus \{t\}$ to the node t, deleting the nodes in $\mathcal{T} \setminus \{t\}$ does not change the max flow from any subset $\mathcal{S}'$ of $\mathcal{S}$ to t and, hence, preserves the min cut, i.e.:

$$m_{\mathcal{S}',t}(\mathbf{c}) = m_{\mathcal{S}',t}(\mathbf{c}_t) \text{ for all } \mathcal{S}' \subseteq \mathcal{S}.$$

For a block length n, let $(f_e^{(n)} : e \in \mathcal{E})$ and $(g_t^{(n)} : t \in \mathcal{T})$ be the encoder and decoder mappings for a rate-$\mathbf{c}$ linear code for the network $(\mathcal{V}, \mathcal{E})$ constructed using the random linear code construction of Theorem 1. Since the code construction relies only on the edge capacities, the mappings $(f_e^{(n)} : e \in \mathcal{E}_t)$ and $g_t^{(n)}$ also define a rate-$\mathbf{c}_t$ random linear code for the network $(\mathcal{V}_t, \mathcal{E}_t)$. Therefore, by Theorem 1, a sufficient condition for the error probability to vanish on $(\mathcal{V}_t, \mathcal{E}_t)$ as n grows without bound is:

$$m_{\mathcal{S}',t}(\mathbf{c}_t) > H(X_{\mathcal{S}'}|X_{\mathcal{S} \setminus \mathcal{S}'}, Y_t) \text{ for all } \mathcal{S}' \subseteq \mathcal{S}.$$

Finally, note that:

$$\Pr(g_t^{(n)}(X_{\mathcal{S}}^n, Y_t^n) \neq X_{\mathcal{S}}^n \text{ for some } t \in \mathcal{T}) \leq \sum_{t \in \mathcal{T}} \Pr(g_t^{(n)}(X_{\mathcal{S}}^n, Y_t^n) \neq X_{\mathcal{S}}^n).$$

Thus, a sufficient condition for the overall error probability to vanish as n grows without bound is:

$$m_{\mathcal{S}',t}(\mathbf{c}_t) > H(X_{\mathcal{S}'}|X_{\mathcal{S} \setminus \mathcal{S}'}, Y_t) \text{ for all } \mathcal{S}' \subseteq \mathcal{S}$$

for every $t \in \mathcal{T}$. Since $m_{\mathcal{S}',t}(\mathbf{c}_t) = m_{\mathcal{S}',t}(\mathbf{c})$, this shows that every rate vector $\mathbf{c} \in \mathcal{R}_C$ is achievable using linear coding. Thus, $\mathcal{R}_C \subseteq \mathcal{R}_L$. ∎

5. PRACTICAL APPROACHES

There have been several works aimed at incorporating network coding and compression. Some of them take an analog approach to coding. This approach is well illustrated by [26], [27]. It is in essence a distortion-based approach, similar to that used in particular for video distribution [28], [29] even in the absence of network coding, that uses simple source/network coding to implement the principles presented in Section 3. Issues are composability of codes in the analog domain, where distortion can be cumulative across nodes.

In the area of lossless multicast, one of the main challenges is to provide a joint source-network code that does not require the high complexity overhead of typical set decoding. One proposed approach is to rely on Turbo codes to build source-network codes which combine good performance with known decoding techniques [30], even though those techniques do not have simple complexity characterization. Another challenge is deciding to what extent redundancy should be whittled down at different sources. Excessive reduction of redundancy will not allow the system to take advantage of what is in effect a virtual mirror system provided by the redundancy of data at sources, and insufficient reduction can lead to transmission of unuseful degrees of freedom. This optimization has been considered when redundancy among sources occurs naturally [31] or is designed [32]. The latter shows the possibility to use network coding to generalize mirror sites, which have heretofore relied on using replication of files.

APPENDIX

In the following we consider the error probability for random vector linear network coding. For simplicity, we state and prove the error probability bound for the case of two sources.

The proof uses the method of types, and is a generalization of the approach of Csiszár [33] for analyzing linear Slepian-Wolf coding. The type $P_{\mathbf{x}}$ of a vector $\mathbf{x} \in \mathbb{F}_2^n$ is the distribution on $\mathbb{F}_2$ defined by the relative frequencies of the elements of $\mathbb{F}_2$ in $\mathbf{x}$, and joint types $P_{\mathbf{xy}}$ are analogously defined.

Theorem 3. *The decoding error probability at each sink is at most $\sum_{i=1}^{3} p_e^i$, where:*

$$
p_e^1 \le \exp \left\{ -n \min_{X,Y} \left(D(P_{XY}||Q) + \left| m_1 (1 - \frac{1}{n} \log L) - H(X|Y) \right|^+ \right) \right.
$$

$$
\left. + 2^{2r_1 + r_2} \log(n+1) \right\}
$$

$$p_e^2 \le \exp\left\{-n\min_{X,Y}\left(D(P_{XY}||Q) + \left|m_2(1 - \frac{1}{n}\log L) - H(Y|X)\right|^+\right)\right.$$

$$\left. + 2^{r_1+2r_2}\log(n+1)\right\}$$

$$p_e^3 \le \exp\left\{-n\min_{X,Y}\left(D(P_{XY}||Q) + \left|m_3(1 - \frac{1}{n}\log L) - H(XY)\right|^+\right)\right.$$

$$\left. + 2^{2r_1+2r_2}\log(n+1)\right\}$$

and X, Y are dummy random variables with joint distribution P_{XY}.

Proof. We provide here an outline of the proof; the full proof can be found in [15].

We consider transmission of the source vector $[\mathbf{x}_1, \mathbf{x}_2] \in \mathbb{F}_2^{n(r_1+r_2)}$. The transfer matrix $\mathbf{C}_{\mathcal{I}(t)}$ specifies the mapping from the source vector $[\mathbf{x}_1, \mathbf{x}_2]$ to the vector $\mathbf{z}$ of bits on the set $\mathcal{I}(t)$ of terminal arcs incident to sink t.

The decoding function at t maps a vector $\mathbf{z}$ of received bits onto a vector $[\tilde{\mathbf{x}}_1, \tilde{\mathbf{x}}_2] \in \mathbb{F}_2^{n(r_1+r_2)}$ that minimizes $\alpha(P_{\mathbf{x}_1\mathbf{x}_2})$ subject to $[\mathbf{x}_1, \mathbf{x}_2]\mathbf{C}_{\mathcal{I}(t)} = \mathbf{z}$. For a minimum entropy decoder, $\alpha(P_{\mathbf{x}_1\mathbf{x}_2}) \equiv H(P_{\mathbf{x}_1\mathbf{x}_2})$, while for a maximum Q-probability decoder, $\alpha(P_{\mathbf{x}_1\mathbf{x}_2}) \equiv -\log Q^n(\mathbf{x}_1\mathbf{x}_2)$. There are three types of decoding errors: in the first type, the decoder outputs the correct value for $\mathbf{x}_2$ but the wrong value for $\mathbf{x}_1$; in the second, the decoder outputs the correct value for $\mathbf{x}_1$ but the wrong value for $\mathbf{x}_2$; in the third, the decoder outputs wrong values for both $\mathbf{x}_1$ and $\mathbf{x}_2$. We obtain upper bounds on the probabilities of these three types of errors, denoted p_e^1, p_e^2, p_e^3 respectively.

(Joint) types of sequences are considered as (joint) distributions P_X ($P_{X,Y}$, etc.) of dummy variables X, Y, etc. The set of different types of sequences in $\mathbb{F}_2^k$ is denoted by $\mathscr{P}(\mathbb{F}_2^k)$. We define the sets of types:

$$\mathscr{P}_n^i = \begin{cases} \{P_{X\tilde{X}Y\tilde{Y}} \in \mathscr{P}(\mathbb{F}_2^{nr_1} \times \mathbb{F}_2^{nr_1} \times \mathbb{F}_2^{nr_2} \times \mathbb{F}_2^{nr_2})|\tilde{X} \ne X, \tilde{Y} = Y\} & i=1 \\ \{P_{X\tilde{X}Y\tilde{Y}} \in \mathscr{P}(\mathbb{F}_2^{nr_1} \times \mathbb{F}_2^{nr_1} \times \mathbb{F}_2^{nr_2} \times \mathbb{F}_2^{nr_2})|\tilde{X} = X, \tilde{Y} \ne Y\} & i=2 \\ \{P_{X\tilde{X}Y\tilde{Y}} \in \mathscr{P}(\mathbb{F}_2^{nr_1} \times \mathbb{F}_2^{nr_1} \times \mathbb{F}_2^{nr_2} \times \mathbb{F}_2^{nr_2})|\tilde{X} \ne X, \tilde{Y} \ne Y\} & i=3, \end{cases}$$

where $\tilde{X} \neq X$ ($\tilde{Y} \neq Y$) in the above definitions designates that the underlying vector that defined the type were not identical. We also define the sets of sequences:

$$\mathscr{T}_{XY} = \{[\mathbf{x}_1, \mathbf{x}_2] \in \mathbb{F}_2^{n(r_1+r_2)} | P_{\mathbf{x}_1\mathbf{x}_2} = P_{XY}\}$$

$$\mathscr{T}_{\tilde{X}\tilde{Y}|XY}(\mathbf{x}_1\mathbf{x}_2) = \{[\tilde{\mathbf{x}}_1, \tilde{\mathbf{x}}_2] \in \mathbb{F}_2^{n(r_1+r_2)} | P_{\tilde{\mathbf{x}}_1\tilde{\mathbf{x}}_2\mathbf{x}_1\mathbf{x}_2} = P_{\tilde{X}\tilde{Y}XY}\}.$$

We can then write the error probability expressions:

$$p_e^1 \leq \sum_{\substack{P_{X\tilde{X}Y\tilde{Y}} \in \mathscr{P}_n^1: \\ \alpha(P_{\tilde{X}\tilde{Y}}) \leq \alpha(P_{XY})}} \sum_{\substack{(\mathbf{x}_1, \mathbf{x}_2) \in \\ \mathscr{T}_{XY}}} Q^n(\mathbf{x}_1\mathbf{x}_2)$$

$$\times \Pr\left(\exists(\tilde{\mathbf{x}}_1, \tilde{\mathbf{x}}_2) \in \mathscr{T}_{\tilde{X}\tilde{Y}|XY}(\mathbf{x}_1\mathbf{x}_2) \text{ s.t.} [\mathbf{x}_1 - \tilde{\mathbf{x}}_1, \mathbf{0}]\mathbf{C}_{\mathcal{I}(t)} = \mathbf{0}\right)$$

$$\leq \sum_{\substack{P_{X\tilde{X}Y\tilde{Y}} \in \mathscr{P}_n^1: \\ \alpha(P_{\tilde{X}\tilde{Y}}) \leq \alpha(P_{XY})}} \sum_{(\mathbf{x}_1, \mathbf{x}_2) \in \mathscr{T}_{XY}} Q^n(\mathbf{x}_1\mathbf{x}_2)$$

$$\times \min\left\{ \sum_{\substack{(\tilde{\mathbf{x}}_1, \tilde{\mathbf{x}}_2) \in \\ \mathscr{T}_{\tilde{X}\tilde{Y}|XY}(\mathbf{x}_1\mathbf{x}_2)}} \Pr\left([\mathbf{x}_1 - \tilde{\mathbf{x}}_1, \mathbf{0}]\mathbf{C}_{\mathcal{I}(t)} = \mathbf{0}\right), 1 \right\}$$

$$p_e^2 \leq \sum_{\substack{P_{X\tilde{X}Y\tilde{Y}} \in \mathscr{P}_n^2: \\ \alpha(P_{X\tilde{Y}}) \leq \alpha(P_{XY})}} \sum_{(\mathbf{x}_1, \mathbf{x}_2) \in \mathscr{T}_{XY}} Q^n(\mathbf{x}_1\mathbf{x}_2)$$

$$\times \min\left\{ \sum_{\substack{(\tilde{\mathbf{x}}_1, \tilde{\mathbf{x}}_2) \in \\ \mathscr{T}_{\tilde{X}\tilde{Y}|XY}(\mathbf{x}_1\mathbf{x}_2)}} \Pr\left([\mathbf{0}, \mathbf{x}_2 - \tilde{\mathbf{x}}_2]\mathbf{C}_{\mathcal{I}(t)} = \mathbf{0}\right), 1 \right\}$$

$$p_e^3 \leq \sum_{\substack{P_{X\tilde{X}Y\tilde{Y}} \in \mathscr{P}_n^3: \\ \alpha(P_{\tilde{X}\tilde{Y}}) \leq \alpha(P_{XY})}} \sum_{(\mathbf{x}_1, \mathbf{x}_2) \in \mathscr{T}_{XY}} Q^n(\mathbf{x}_1\mathbf{x}_2)$$

$$\times \min\left\{ \sum_{\substack{(\tilde{\mathbf{x}}_1, \tilde{\mathbf{x}}_2) \in \\ \mathscr{T}_{\tilde{X}\tilde{Y}|XY}(\mathbf{x}_1\mathbf{x}_2)}} \Pr\left([\mathbf{x}_1 - \tilde{\mathbf{x}}_1, \mathbf{x}_2 - \tilde{\mathbf{x}}_2]\mathbf{C}_{\mathcal{I}(t)} = \mathbf{0}\right), 1 \right\},$$

where the probabilities are taken over realizations of the network transfer matrix $\mathbf{C}_{\mathcal{I}(t)}$ corresponding to the random network code.

We can use simple cardinality bounds on $|\mathscr{P}_n^i|, |\mathscr{T}_{XY}|$ and $|\mathscr{T}_{\tilde{X}\tilde{Y}|XY}(\mathbf{x}_1\mathbf{x}_2)|$, along with the identity:

$$Q^n(\mathbf{x}_1\mathbf{x}_2) = \exp\{-n(D(P_{XY}\|Q) + H(XY))\},$$

$$(\mathbf{x}_1, \mathbf{x}_2) \in \mathscr{T}_{XY},$$

to obtain:

$$
p_e^1 \le \exp\left\{ -n \min_{\substack{P_{X\tilde{X}Y\tilde{Y}} \in \mathscr{P}_n^1: \\ \alpha(P_{\tilde{X}Y}) \le \alpha(P_{XY})}} \left(D(P_{XY}\|Q) + \left| -\frac{1}{n}\log P_1 - H(\tilde{X}|XY) \right|^+ \right) \right.
$$

$$
\left. + 2^{2r_1 + r_2} \log(n+1) \right\}
$$

$$
p_e^2 \le \exp\left\{ -n \min_{\substack{P_{X\tilde{X}Y\tilde{Y}} \\ \in \mathscr{P}_n^2: \\ \alpha(P_{X\tilde{Y}}) \le \\ \alpha(P_{XY})}} \left(D(P_{XY}\|Q) + \left| -\frac{1}{n}\log P_2 - H(\tilde{Y}|XY) \right|^+ \right) \right.
$$

$$
\left. + 2^{r_1 + 2r_2} \log(n+1) \right\}
$$

$$
p_e^3 \le \exp\left\{ -n \min_{\substack{P_{X\tilde{X}Y\tilde{Y}} \\ \in \mathscr{P}_n^3: \\ \alpha(P_{\tilde{X}\tilde{Y}}) \le \\ \alpha(P_{XY})}} \left(D(P_{XY}\|Q) + \left| -\frac{1}{n}\log P_3 - H(\tilde{X}\tilde{Y}|XY) \right|^+ \right) \right.
$$

$$
\left. + 2^{2r_1 + 2r_2} \log(n+1) \right\},
$$

where the exponents and logs are taken with respect to base 2, and:

$$P_1 = \Pr\left([\mathbf{x}_1 - \tilde{\mathbf{x}}_1, \, 0]\mathbf{C}_{\mathcal{I}(t)} = 0\right)$$

$$P_2 = \Pr\left([0, \mathbf{x}_2 - \tilde{\mathbf{x}}_2]\mathbf{C}_{\mathcal{I}(t)} = 0\right)$$

$$P_3 = \Pr\left([\mathbf{x}_1 - \tilde{\mathbf{x}}_1, \mathbf{x}_2 - \tilde{\mathbf{x}}_2]\mathbf{C}_{\mathcal{I}(t)} = 0\right), \quad \mathbf{x}_1 - \tilde{\mathbf{x}}_1, \mathbf{x}_2 - \tilde{\mathbf{x}}_2 \neq 0.$$

P_1, P_2, and P_3 represent the probabilities that two distinct source sequences (differing in the value of one or both source nodes) are mapped by the network code to the same observed sequence at the sink; note that with a random linear network code, the probabilities are unchanged for any non-zero $\mathbf{x}_1 - \tilde{\mathbf{x}}_1, \mathbf{x}_2 - \tilde{\mathbf{x}}_2$. These probabilities can be bounded in terms of n and the network parameters $m_i, i = 1, 2$, the minimum cut capacity between the receiver and source X_i, m_3, the minimum cut capacity between the receiver and both sources, and L, the maximum source-receiver path length. In particular, we can show by induction that:

$$P_i \leq \left(1 - \left(1 - \frac{1}{2^n}\right)^L\right)^{m_i}$$

$$\leq \left(\frac{L}{2^n}\right)^{m_i}. \tag{2}$$

For the minimum entropy decoder, we have:

$$\alpha(P_{\tilde{X}\tilde{Y}}) \leq \alpha(P_{XY}) \Rightarrow \begin{cases} H(\tilde{X}|XY) \leq H(\tilde{X}|Y) \leq H(X|Y) & \text{for } Y = \tilde{Y} \\ H(\tilde{Y}|XY) \leq H(\tilde{Y}|X) \leq H(Y|X) & \text{for } X = \tilde{X} \\ H(\tilde{X}\tilde{Y}|XY) \leq H(\tilde{X}\tilde{Y}) \leq H(XY) \end{cases}$$

which gives:

$$p_e^1 \leq \exp\left\{-n \min_{XY}\left(D(P_{XY}||Q) + \left|-\frac{1}{n}\log P_1 - H(X|Y)\right|^+\right)\right.$$

$$\left. + 2^{2r_1 + r_2}\log(n+1)\right\}$$

$$p_e^2 \leq \exp\left\{-n\min_{XY}\left(D(P_{XY}||Q) + \left|-\frac{1}{n}\log P_2 - H(Y|X)\right|^+\right)\right.$$

$$\left. + 2^{r_1+2r_2}\log(n+1)\right\}$$

$$p_e^3 \leq \exp\left\{-n\min_{XY}\left(D(P_{XY}||Q) + \left|-\frac{1}{n}\log P_3 - H(XY)\right|^+\right)\right.$$

$$\left. + 2^{2r_1+2r_2}\log(n+1)\right\}.$$

It can also be shown that the same bounds hold for the maximum Q-probability decoder. Combining these bounds with the bounds on P_i from Equation (2) gives the result. ∎

REFERENCES

[1] D. Slepian and J. K. Wolf. Noiseless coding of correlated information sources. *IEEE Transactions on Information Theory*, 19: 471–480, 1973.

[2] R. Ahlswede and J. Körner. Source coding with side information and a converse for degraded broadcast channels. *IEEE Transactions on Information Theory*, 21: 629–637, Nov. 1975.

[3] H. Yamamoto. Wyner-Ziv theory for a general function of the correlated sources. *IEEE Transactions on Information Theory*, 28: 1788–1791, Sept. 1982.

[4] A. Orlitsky and J. R. Roche. Coding for computing. *IEEE Transactions on Information Theory*, 47: 903–917, Mar. 2001.

[5] H. S. Witsenhausen. The zero-error side information problem and chromatic numbers. *IEEE Transactions on Information Theory*, 22: 592–593, 1976.

[6] N. Alon and A. Orlitsky. Source coding and graph entropies. *IEEE Transactions on Information Theory*, 42: 1329–1339, Sept. 1996.

[7] A. D. Wyner and J. Ziv. The rate-distortion function for source coding with side information at the decoder. *IEEE Transactions on Information Theory*, 22: 1–10, Jan. 1976.

[8] T. Berger and S. Y. Tung. Encoding of correlated analog sources. In *Proceedings of IEEE-USSR Joint Workshop on Information Theory*, pp. 7–10, 1975.

[9] R. M. Gray and A. D. Wyner. Source coding for a simple network. *Bell Systems Technical Journal*, 53: 1681–1721, Nov. 1974.

[10] A. D. Wyner. On source coding with side information at the decoder. *IEEE Transactions on Information Theory*, 21: 294–300, Nov. 1975.

[11] T. Berger and R. Yeung. Multiterminal source encoding with one distortion criterion. *IEEE Transactions on Information Theory*, 35(2): 228–236, 1989.

[12] R. Ahlswede, N. Cai, S.-Y. R. Li, and R. W. Yeung. Network information flow. *IEEE Transactions on Information Theory*, 46: 1204–1216, July 2000.

[13] R. Koetter and M. Médard. An algebraic approach to network coding. *IEEE/ACM Transactions on Networking*, 11: 782–795, Oct. 2003.

[14]   L. Song, R. W. Yeung, and N. Cai. Zero-error network coding for acyclic networks. *IEEE Transactions on Information Theory*, 49: 3129–3139, July 2003.

[15]   T. Ho, M. Médard, R. Koetter, D. R. Karger, M. Effros, J. Shi, and B. Leong. A random linear network coding approach to multicast. *IEEE Transactions on Information Theory*, 52: 4413–4430, Oct. 2006.

[16]   M. Effros, M. Médard, T. Ho, S. Ray, D. Karger, and R. Koetter. Linear network codes: A unified framework for source, channel, and network coding. In *Proceedings of the DIMACS Workshop on Network Information Theory*, (Piscataway, NJ), IEEE, Mar. 2003. Invited paper.

[17]   A. Ramamoorthy, K. Jain, P. A. Chou, and M. Effros. Separating distributed source coding from network coding. *Joint Special Issue of the IEEE Transactions on Information Theory and the IEEE/ACM Transactions on Networking*, 52: 2785–2795, June 2006.

[18]   T. Berger. Multiterminal source coding. In *The Information Theory Approach to Communications*, (New York), pp. 172–231, CISM Courses and Lectures, Springer-Verlag, July 1977.

[19]   A. R. Lehman and E. Lehman. Complexity classifications of network information flow problems. In *Allerton Annual Conference on Communications, Control, and Computing*, (Monticello, IL), Sept. 2003.

[20]   M. Langberg, M. Sprintson, and J. Bruck. Network coding: A computational perspective. *IEEE Transactions on Information Theory*, 55(1): 145–157, 2008.

[21]   H. Yao and E. Verbin. Network coding is highly non-approximable. In *Allerton Annual Conference on Communications, Control, and Computing*, 2009.

[22]   W.-H. Gu. *On achievable rate regions for source coding networks*. Ph.D. dissertation, California Institute of Technology, Pasadena, CA, 2009.

[23]   W.-H. Gu and M. Effros. On the concavity of rate regions for lossless source coding in networks. In *Proceedings of the IEEE International Symposium on Information Theory*, (Seattle, WA), pp. 1599–1603, July 2006.

[24]   D. Slepian and J. K. Wolf. A coding theorem for multiple access channels with correlated information sources. *Bell System Technical Journal*, 52: 1037–1076, 1973.

[25]   T. M. Cover and J. A. Thomas. *Elements of Information Theory*. Wiley, second ed., 2006.

[26]   S. Katti, S. Shintre, S. Jaggi, D. Katabi, and M. Médard. Real network codes breaking the all-or-nothing barrier. In *Allerton Annual Conference on Communications, Control, and Computing*, 2007.

[27]   B. Dey, S. Katti, S. Jaggi, D. Katabi, and M. Médard. "Real" and "complex" network codes—promises and challenges. In *Fourth Workshop on Network Coding Theory and Applications (NETCOD)*, 2008.

[28]   G. Woo, D. Katabi, and S. Chachulski. One video stream to serve diverse receivers. Tech. rep., MIT, 2008.

[29]   D. Katabi, R. Hariharan, and S. Jakubczak. Softcast: One video to serve all wireless receivers. Mit-csail-tr-2009-005, MIT, 2009.

[30]   G. Maierbacher, J. Barros, and M. Médard. Practical source-network decoding. In *IEEE International Symposium on Wireless Communication Systems*, 09.

[31]   A. Lee, M. Médard, K. Z. Haigh, S. Gowan, and P. Rubel. Minimum-cost subgraphs for joint distributed source and network coding. In *Third Workshop on Network Coding, Theory, and Applications*, 2007.

[32]   S. Huang, A. Ramamoorthy, and M. Médard. Minimum cost mirror sites using network coding: Replication vs. coding at the source nodes. *IEEE Transactions on Information Theory*, 57(2): 2011.

[33]   I. Csiszár. Linear codes for sources and source networks: Error exponents. *IEEE Transactions on Information Theory*, 28: 585–592, July 1982.

CHAPTER 9

Scaling Laws with Network Coding

Atilla Eryilmaz[1] and Lei Ying[2]

[1]Department of Electrical and Computer Engineering, Ohio State University, Columbus, OH, USA;
[2]Department of Electrical and Computer Engineering, Iowa State University, Ames, IA, USA

Contents

Abstract

In this chapter, we overview the scaling law results associated with network coding as the network size and/or the coding window size scale. These results highlight the potential gains to be derived from network coding in various scenarios of interest. Our discussion covers the scaling of throughput and delay performance metrics for wireless networks both under time-varying channel conditions and node mobility.

Keywords: Scaling laws, random network coding, throughput-delay trade-offs, mobility models, mobile *ad hoc* networks, wireless scheduling.

Network Coding. DOI: 10.1016/B978-0-12-380918-6.00009-3
Copyright © 2012 Elsevier Inc. All rights reserved.

1. INTRODUCTION AND BASIC SETUP

Thanks to the framework established by Claude E. Shannon in [27], coding has long been acknowledged to provide throughput gains over communication links. The fundamental strength of coding is its ability to spread information across long chunks of transmission signals to achieve robustness against statistical channel variations. While this idea has transformed the world by enabling a global communication network, most of the exciting advances have been limited to point-to-point links with a few other capacity results relating to limited multi-user scenarios. Within the last decade, the idea of *network coding* has emerged [2, 17, 19], suggesting a means of achieving significant throughput gains in general networks through simple algebraic coding strategies. In this article, we are interested in the scaling laws associated with the throughput and delay performance of communication systems when they can employ coding, in particular network coding, strategies. These scaling laws and their comparison to non-coding techniques help quantify the gains that coding can provide to large-scale systems.

In our discussion, we consider packetized communication networks, where each *packet*, denoted as $\mathbf{P}$, is represented by an m dimensional vector over a finite field $\mathbb{F}_q$. Due to the algebraic nature of this construction, it is possible to perform algebraic operations among different packets (i.e., vectors) to obtain a new packet in the vector space. Such ability yields the (linear) network coding capability of the network. Throughout our discussion, we will mostly assume time-slotted system operation, where all nodes are synchronized to a common clock and operate in fixed duration time slots. With this setup, we define the network coding operation more rigorously next.

Definition 1 ((Linear) Network Coding). (Linear) Network Coding *refers to the mode of transmission of any network node where in slot t, any linear combination of a set $\{\mathbf{P}_k\}_{k \geq 1}$ of available packets can be formed. Specifically, we have:*

$$\mathbf{P}[t] = \sum_{k \geq 1} a_k[t]\mathbf{P}_k, \tag{1}$$

where $a_k[t] \in \mathbb{F}_q$, for each k constitute the global coding coefficients.

In comparison, we also define another extreme operating mode where each node is allowed to transmit one of its available packets. This is called the scheduling mode, and is made more precise next.

Definition 2 (Scheduling). Scheduling *refers to the traditional mode of transmission without network coding where, in any given slot t, the transmitter must pick for transmission a single packet from the available set of packets, $\{\mathbf{P}_k\}_{k \geq 1}$. Specifically, we have $\mathbf{P}[t] \in \{\mathbf{P}_k : k = 1, 2, \ldots\}$.*

We start by noting that the scenario of a transmitter broadcasting data to all (or a subset of) receivers over unreliable channels forms the building block of multicast communication in cellular as well as in multi-hop wireless networks. The erasure channels may be used to model either physical channels as in cellular or satellite communication, or virtual channels representing end-to-end transmissions through multiple cascaded transmissions as in multi-hop wireless networks. We will divide our discussion to two scenarios whereby we first study scaling laws associated with the static network scenario in Section 2, and later discuss the mobile *ad hoc* network scenario in Section 3.

2. WIRELESS BROADCAST OVER LOSSY LINKS

In this section, we focus on tight bounds for the throughput and delay scaling laws associated with a range of wireless broadcast scenarios over lossy links. Our discussion under this item is divided into three parts: (1) delay scaling, where we provide the insights and the main results on the delay improvements achieved by network coding; (2) extensions, where we build on the first part to consider extensions in the network topology, arrival dynamics, and application sensitivities to delay; and, finally, (3) throughput-delay trade-offs, where we reveal the throughput deficiencies of earlier coding results, and provide recent results on the delay versus throughput scaling that is achievable by network coding.

2.1. Delay Scaling Gains

Figure 9.1 captures the related download scenario introduced and analyzed in [10, 11], where K packets are to be broadcast over erasure channels to all N receivers. To capture the channel variations, albeit crudely, it is assumed that the channel state process of the nth user, denoted $\{C_n[t]\}_{t \geq 0}$,

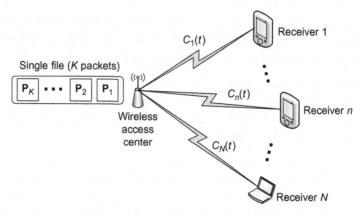

Figure 9.1 Rateless broadcast over N independent erasure channels, where $C_n[t]$ is Bernoulli distributed with mean c_n.

is a Bernoulli Process with mean $c_n \in [0, 1]$ that is independent across users. The nth receiver successfully receives the packet $\mathbf{P}[t]$ transmitted in slot t if, and only if, $C_n[t] = 1$.

The first question to answer in this scenario is whether network coding can outperform the optimal scheduling policy, even when the transmitter has perfect Channel-Side-Information (CSI) at the outset of its transmission. Here, CSI implies the knowledge of the realization $\mathbf{C}[t] \triangleq (C_n[t])_{n=1}^{N}$ at the beginning of the time slot t, for any t. The following example, provided in [10], proves that such a gain is possible starting from $K = 3$ and $N = 3$.[1]

Example 1. *Consider the case of $K = 3$ and $N = 3$, i.e., three packets are to be broadcast to three receivers. Consider the channel realizations $\mathbf{C}[1] = (0, 1, 1), \mathbf{C}[2] = (1, 0, 1), \mathbf{C}[3] = (1, 1, 0),$ and $\mathbf{C}[4] = (1, 1, 1)$. Thus, in the first four slots, each receiver can hear the transmission three times. The optimal scheduling rule would transmit $\mathbf{P}_1, \mathbf{P}_2, \mathbf{P}_3$ in the first three slots, leaving Receiver i in demand for Packet i in the fourth slot. Clearly, no scheduling rule can ever*

---

[1] We note that for $N = 2$, it is possible to devise a scheduling strategy that can achieve the same performance as network coding strategy *when CSI is available at the outset of transmissions*. However, we shall see that network coding can asymptotically achieve the same performance without CSI.

Table 9.1 Demonstration of *Example 1*: R_i corresponds to Receiver i, "−" denotes OFF channel states, and the entry a|b gives the optimal transmissions with scheduling | coding, respectively. With scheduling, no choice of $\{P_i\}$ in slot 4 can complete the file at all the receivers for the given channel realization

	$t = 1$	$t = 2$	$t = 3$	$t = 4$
R_1	−	$\mathbf{P}_2\|\mathbf{P}_2$	$\mathbf{P}_3\|\mathbf{P}_3$	?|$(\mathbf{P}_1+\mathbf{P}_2+\mathbf{P}_3)$
R_2	$\mathbf{P}_1\|\mathbf{P}_1$	−	$\mathbf{P}_3\|\mathbf{P}_3$	?|$(\mathbf{P}_1+\mathbf{P}_2+\mathbf{P}_3)$
R_3	$\mathbf{P}_1\|\mathbf{P}_2$	$\mathbf{P}_2\|\mathbf{P}_2$	−	?|$(\mathbf{P}_1+\mathbf{P}_2+\mathbf{P}_3)$

complete the file download at all three receivers in the fourth slot since each receiver misses a different packet. With coding, on the other hand, the following transmissions will complete the transmissions in four slots: $\mathbf{P}_1, \mathbf{P}_2, \mathbf{P}_3, (\mathbf{P}_1 + \mathbf{P}_2 + \mathbf{P}_3)$ *(see Table 9.1). It is not difficult to see that coding will never require more slots than is necessary for scheduling for all other realizations. Hence, we achieve strictly better completion times with coding.*

Example 1 shows the existence of channel realizations leading to strict performance improvement of network coding over any scheduling strategy. Yet, it is unclear whether such an advantage leads to any significant improvements in the mean completion time as the system size scales. This problem is studied in several works ([10, 12, 24]). Here, we provide the main results on the asymptotic *mean completion time* of a block of size K packets under the following *Random Network Coding (RNC)* strategy.

RANDOM NETWORK CODING (RNC):

While (All K packets are not decoded at all N receivers)
    Pick $a_k[t]$ uniformly at random from $\mathbb{F}_q$ for each k;
    Transmit $\mathbf{P}[t] = \sum_{k=1}^{K} a_k[t]\mathbf{P}_k$;
    $t \leftarrow t + 1$;

Theorem 1. *Let* $Z^{RNC}(N, K, \mathbf{c})$ *denote the completion time of RNC of a block of K packets to N receivers over independently varying erasure channels with mean vector* $\mathbf{c} \triangleq (c_n)_{n=1}^{N}$. *Then, the above RNC algorithm is asymptotically optimal as the field size q tends to infinity in the sense that the mean completion time* *is minimized over all other policies, and converges to:*

$$\mathbb{E}[Z^{RNC}(N, K, \mathbf{c})] =$$

$$K + \sum_{t=K}^{\infty} \left[1 - \prod_{i=1}^{N} \left(\sum_{j=K}^{t} \binom{j-1}{K-1} (1 - c_i)^{(j-K)} c_i^K \right) \right].$$

Moreover, the first and second moments of $Z^{RNC}(N, K, \mathbf{c})$, for fixed K and increasing N takes the following form under symmetric channel conditions, i.e., when $c_n = c \in (0, 1)$ for all $n = 1, \ldots, N$: let $\mathrm{lc}(\cdot)$ be a shorthand for $\log_{\frac{1}{c}}(\cdot)$. Then, we have [11]:

$$\mathbb{E}[Z^{RNC}(N, K, c)] = \mathrm{lc}(T) + \frac{1}{2} - \frac{\gamma}{\log(1-c)} + h(\mathrm{lc}(T)) + o(1),$$

$$\mathbb{E}[(Z^{RNC}(N, K, c))^2] = \mathrm{lc}^2(T) + \mathrm{lc}(T)(1 + 2\gamma + 2g_1(\mathrm{lc}(T))) + \frac{2}{3}$$

$$- \frac{\gamma}{\log(1-c)} - \frac{(\gamma^2 + (\pi^2/6))}{\log^2(1-c)}$$

$$+ O((K-1)\mathrm{lc}(\mathrm{lc}(N))) + h(\mathrm{lc}(T))$$

$$+ g_2(\mathrm{lc}(T)) + o(1),$$

where $T = N \left(\frac{c}{(1-c)} \right)^{K-1} \frac{\mathrm{lc}^{(K-1)} N}{(K-1)!}$, and γ is the Euler-Mascheroni constant (approximately equal to 0.5772), and $h(\cdot)$ is a periodic C^{∞}-function[2] of period 1 and mean 0 with Fourier coefficients $\hat{h}_k = \frac{1}{\log(c)}\Gamma(\frac{2ik\pi}{\log(c)})$, for $k = 0, 1, \ldots$, where $\Gamma(\cdot)$ is the standard gamma function. Similarly, $g_1(\cdot)$, and $g_2(\cdot)$ are two periodic C^{∞}-functions of period 1 and mean 0.

The key result to take from the last two asymptotic expressions is simple: $\mathbb{E}[Z^{RNC}(N, K, c)] \approx \mathrm{lc}\, N$ and $\mathbb{E}[(Z^{RNC}(N, K, c))^2] \approx \mathrm{lc}^2\, N$ for fixed (K, c) and increasing N. Since Jensen's inequality requires that $\mathbb{E}[(Z^{RNC}(N, K, c))^2] \geq \mathbb{E}[Z^{RNC}(N, K, c)]^2$, these results imply the asymptotic second-moment optimality of RNC as N tends to infinity.

---

[2] A periodic C^{∞}-function $h(x)$ is an infinitely continuously differentiable function for which there exists a constant P (period) such that $h(x + P) = h(x)$ for all x.

In a more recent work, the convergence of the completion time distribution is also studied in [33] using *extreme value theory*. Extreme value theory studies the characteristics of random variables of the form $Z_N = \max_{1 \leq n \leq N} Y_n$, where Y_n are independent and identically distributed (i.i.d.) random variables. Then, it is true, under mild assumptions, that the distribution of Z_N converges, as N tends to infinity, to one of three possible extreme value distributions, namely Fréchet, Gumbel, or Weibull ([5, 25]).

Under the operation of RNC, if Y_n is defined as the number of slots required for the nth channel to be ON for exactly K times, then Z_N corresponds to the previously defined completion time $Z^{RNC}(N, K, c)$ of all K packets at all N receivers. It is easy to see that, under the i.i.d. Bernoulli channel processes with mean c, Y_n is a *negative binomial* or a *Pascal* variable of order K and success probability c. Thus, the natural use of extreme value theory leads to the following asymptotic convergence result, established in [33]:

Theorem 2. *For fixed K, and symmetric channel conditions of $c_n = c$ for all n, the completion time $Z^{RNC}(N, K, c) =: Z(N)$ to disseminate K packets to N nodes using RNC-type broadcasting is bounded by random variables belonging to Gumbel domain of attraction ([5, 25]). Namely, there exist $Z_l(N)$ and $Z_u(N) = Z_l(N) + 1$, satisfying:*
(i)

$$Z_l(N) \leq_{st} Z(N) \leq_{st} Z_u(N),$$

where $U \leq_{st} V$ for two random variables implies that $\mathbb{P}(U > a) \geq \mathbb{P}(V > a)$ for all a:
(ii)

$$\sum_{N \to \infty} \mathbb{P}\left(\frac{Z_l(N) - b_N}{a_N} \leq x\right) = \exp(-e^{-x}), \quad \text{for all } x \in \mathbb{R}, \quad (2)$$

where $a_N = -1/\log(c)$, and $b_N = \text{lc}(N) + (K-1)(\text{lc}(\zeta) + \text{lc}((1-c)/c)) - \text{lc}((K-1)!)$, with $\zeta = \text{lc}(N) + (K-1)\text{lc}((1-c)/c)$.

This result implies that the completion time cumulative distribution has a sharp transition around b_N from zero to one, as captured by the limiting doubly exponential Gumbel-type distribution in (2). This result confirms through a different asymptotic that the mean completion time increases as $lc(N)$ as N scales. We note that [33] also studies the more general asymmetric channel state processes.

All of these works fall into the context of *rateless transmission* since they focus on minimizing the transmission time of finite size data. A directly related metric under a rateless transmission scenario, studied in [12, 24], is termed *reliability gain*, which measures the expected number of transmissions required per data packet. With a network coding policy over K packets, this is computed as the completion time of the whole block divided by K. We will overview some of the results on this metric in the next section.

2.2. Extensions

2.2.1 Topological Extensions

The building block of a single-hop network can be extended to more general topologies in several ways. In this section, we discuss two such possibilities that have been explored in the literature. The first topology, introduced and studied in [12], is that of a full N-ary tree of depth h rooted at the source node, where the source aims to transmit a set of K packets to all the N^h leaves of the tree over independently fading erasure links, each with the uniform ON probability of $c \in (0, 1)$ (see Fig. 9.2). The following theorem summarizes its main results in terms of the reliability gain metric introduced at the end of the previous section (we refer the reader to [12] for further details.)

Theorem 3. *For the tree topology of Fig. 9.2 of depth h, consider four schemes for comparison: (i) end-to-end ARQ (e2e-ARQ): the root of the tree transmits packets until all receivers at the leaves receive each packet. The relays only forward the packets they receive once. (ii) link-by-link ARQ (ll-ARQ): the root and each relay make sure the packet is successfully received by all its children. Thus, relays may make multiple transmissions for each packet. (iii) end-to-end FEC (e2e-FEC): the root is allowed to perform block coding to recover the failure of some its transmissions. Relays only forward the packets without further processing. (iv) random network coding (RNC): after successfully decoding all K packets received from its parent, each relay performs random network coding over them (as defined in the RNC Algorithm of Section 2) to guarantee successful reception of its packets at each of its children.*

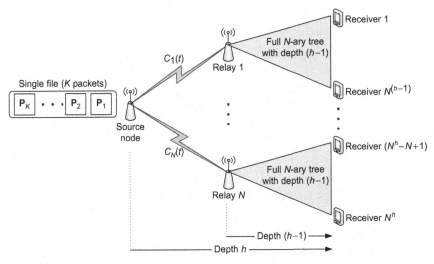

Figure 9.2 Extension of the setup of Fig. 9.1 to a full tree of depth h, where each erasure link fades independently with the uniform mean of c.

Let the relative reliability gain, ρ, of different schemes with respect to network coding be the ratio of reliability gain (number of transmissions for each data packet) of the corresponding scheme to the reliability gain of network coding (the fourth scheme above). Then, the following scaling laws hold for fixed K and increasing N:

- $\rho^{(e2e-ARQ)}(N, c, h) = \Theta(\log N)$.
- $\rho^{(ll-ARQ)}(N, c, h) = \Theta(c^{-(h-1)} \log N)$.
- $\rho^{(e2e-FEC)}(N, c, h) = \Theta(c^{-h})$.

Theorem 3 shows that the relative gains in the number of transmissions scale up in the size of the network as well as the depth of the tree.

Another topological extension, introduced in [11], enables the use of the above single-hop model in multi-hop wireless networks. This is achieved by rearranging the general topology into a layered topology, and then analyzing the layered topology as a chain of single-hop networks. The following example demonstrates the proposed layering approach.

Example 2 (Decomposition of a Network into Layers). *Consider the multicast setting shown in Fig. 9.3 consisting of two sink nodes, a single source node and some intermediate nodes.*

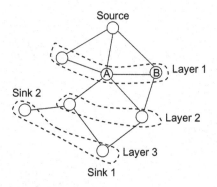

Figure 9.3 A multicast setting in a general network topology.

We decompose the network into layers such that a node belongs to Layer i if the shortest path from the source to it is i hops. We can identify the layer in which each node is to be placed by simply flooding the network or by using shortest path algorithms. We assume a primary interference model that captures CDMA networks whereby a node can only transmit or receive in each slot, but not both. The files generated at the source are transmitted from one layer to the next, subject to such interference constraints. Then, transmission across two subsequent layers i and i + 1 can be obtained by studying an M × R switch,[3] where M and R are the number of nodes in layers i and i + 1, respectively.

Using this layering, an algorithm is proposed in [11] whereby odd and even numbered layers alternate for transmission of their encoded packets in a random access fashion. This leads to a suboptimal but fully distributed multi-hop transmission strategy that extends the single-hop algorithm of Section 2. We refer the interested reader to [11] for further analysis and discussion of this algorithm.

2.2.2 Arrival Dynamics

In this section, we extend the setup from a rateless transmission scenario to a scenario with a stream of incoming data according to some random process to be broadcast to N users over i.i.d. erasure channels with uniform ON probabilities of c (see Fig. 9.4). Data enters the transmitting node in

---

[3] An $M \times R$ network switch forwards the stream of packets arriving at its M input ports to their corresponding R output ports. It is usually represented as a bi-partite graph with directed links from M input ports to each of the R output ports.

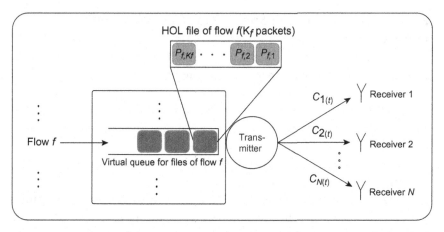

Figure 9.4 Extension of the setup to streaming flows, where HOL stands for Head-of-Line.

groups of K packets, which will be called a *file* in this context. We assume that the files arrive according to a Poisson process with rate λ.

Then, the metric of *waiting time*, $W(N, c)$, of the incoming files before successful broadcast is investigated in [11] for fixed file sizes K, and increasing number of users N. This analysis yields the following result for RNC performance.

Theorem 4. *The waiting time performance of RNC with fixed K and scaling N under symmetric channel conditions is:*

$$W^{RNC}(N, c) = \Theta\left(\frac{\lambda \, lc^2 \, N}{2(1 - \lambda \, lc \, N)}\right), \quad for \; \lambda < \frac{1}{lc \, N}.$$

Moreover, RNC policy is asymptotically optimal in N for minimizing the mean waiting time.

This result extends the mean completion time optimality of RNC of rateless transmission to mean waiting time optimality of RNC for streaming file broadcasts by the transmitter. This extension builds on the asymptotic second-moment optimality of RNC stated after Theorem 1. However, it should be noted that these results are contingent on the fixed K assumption, and therefore the supportable rate λ must decrease as $\frac{1}{lc(N)}$ with increasing N. The effect of the scaling of K will be discussed in Section 2.3.

2.2.3 Accounting for Delay Sensitivities of Incoming Traffic

In this extension, studied in [3], the setup of Fig. 9.4 is extended to the scenario depicted in Fig. 9.5, where, as before, files arrive according to a Poisson process of rate γ, but also associated with an independently generated random variable U that measures its value. Moreover, a price p is paid to the transmitter by each receiver once the file is transmitted to it. The variable U measures the amount of waiting time plus the total transmission cost that the file can accept. Those files with smaller U values are probabilistically dropped as they enter the system, therefore reducing the admitted files to Poisson with a smaller rate of λ.

Clearly, as the price p of each transmission increases, the ratio of admitted packets decreases. On the other hand, as the price reduces, a larger fraction is accepted whereas their waiting time increases. In [3], the problem of setting prices for revenue maximization is studied in terms of changing N and K. Figure 9.6 depicts the maximum revenue performance under RNC transmission strategy as functions of K and N. An interesting observation from this figure is the unimodal nature of the optimum revenue level as a function K. Moreover, it is observable that the choice of optimum K_{opt} is highly insensitive to the size of the network. This is quite attractive as it suggests a natural decoupling between the optimal coding window size and the number of users.

To show the relative gains of network coding compared to traditional scheduling, Fig. 9.7 depicts the performance comparison of RNC with Optimal Scheduling strategy when no CSI is available for a range of K and fixed N. This shows the orders of magnitude improvement in revenue gains that is achievable with a network coding transmission strategy.

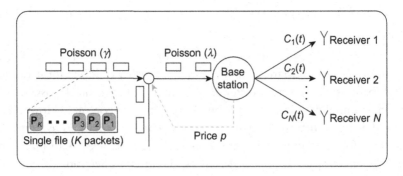

Figure 9.5 Extension of the setup to streaming flows with delay sensitivities.

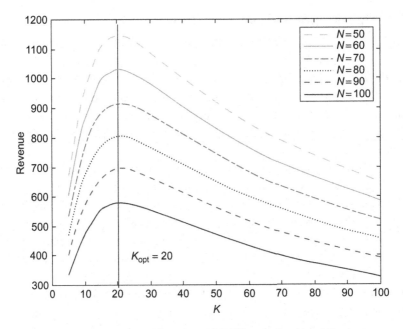

Figure 9.6 Optimum revenue performance of RNC for varying N and K.

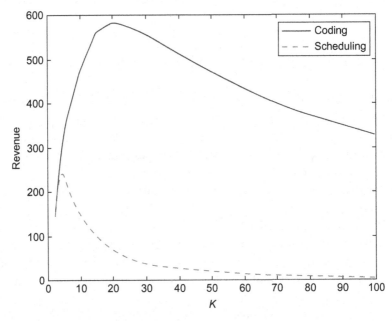

Figure 9.7 Optimum revenue performance comparison of RNC versus Scheduling for varying K with $N = 50$.

2.3. Throughput and Delay Trade-off

All of the aforementioned results focus on the scaling of the completion time and its ramifications with increasing number of receivers, N, dominantly assuming a fixed (coding) block of K packets used in the encoding procedure. However, this only provides limited, and potentially misleading, information about the throughput performance when an infinite amount of data is to be transmitted in blocks. *In fact, it can be seen that when the coding window size is held constant, the throughput of the system goes to zero as the system becomes large.* To clarify this, we let $\tau^{RNC}(N, K, c)$ denote the throughput of RNC when an infinite amount of information at the source is broadcast to N receivers over i.i.d. Bernoulli(c) channels in *blocks of K packets*. To understand the behavior of $\tau^{RNC}(N, K, c)$, we first observe that the completion time of each block of K packets is an independent random variable $Z^{RNC}(N, K, c)$. The block transmission structure, together with the independence of channel states across time, allows us to model the RNC operation as a renewal process with renewals at the start of each coding block formation. Hence, by defining a constant reward of K acquired in each renewal interval, we can utilize the main result from renewal theory [26] to write:

$$\tau^{RNC}(N, K, c) = \frac{K}{\mathbb{E}[Z^{RNC}(N, K, c)]}.$$

Clearly, for any fixed K, $\tau^{RNC}(N, K, c) \downarrow 0$, as N tends to ∞ since $\mathbb{E}[Z^{RNC}(N, K, c)]$ scales as $\mathrm{lc}(N)$ (see Theorem 1). This implies that as the network size N scales, the block size K must also be scaled to achieve non-vanishing throughput levels. Yet, the scaling of K implies scaling of the transmission duration of each block. This motivates the investigation of the proper scaling of K with N that results in the best throughput-delay trade-off.

We start by noting that $\tau^{\pi}(N, K, c) \leq c$ for any policy π since the right-hand-side corresponds to the mean fraction of time each receiver will successfully receive a packet. This upper-bound is achievable only if almost all receptions lead to a useful packet for each receiver. Interestingly, it will be shown that RNC can achieve this bound if K and N assume a certain scaling behavior. The following example highlights the effect of coding window size scaling on the throughput-delay performance of RNC, and motivates the subsequent study.

Example 3. *Consider a single source broadcasting a block of K packets to N receivers using RNC policy over i.i.d. erasure channels with mean c. Using standard random linear coding arguments (e.g., [15]), for a large enough field size d, it is sufficient for the receivers to receive approximately K coded packets to be able to decode the block.*

Each receiver requires at least K slots to decode the whole block. Let the random variable M[t] represent the number of receivers that have successfully decoded K packets in $t \geq K$ time slots. Let r[t] be the probability that any given receiver receives at least K packets in $t \geq K$ time slots. Then, M[t] is a binomial random variable with probability of success r[t], where $r[t] = \sum_{l=K}^{t} \binom{t}{l} c^l (1-c)^{t-l}$. Then $E(M[t]) = Nr[t]$. Thus, r[t] also represents the fraction of receivers that have successfully decoded K packets by t slots.

To compare the behavior of r[t] as a function of t for different values of K, we define a normalized time variable, $s = \frac{t-K}{K}$. Accordingly, we define $r'[s] = r[Ks + K]$, which can be interpreted as the fraction of receivers that have successfully decoded a single packet in a block of K packets by s time slots. The comparison of r'[s] for different K allows us to see, in a normalized time scale, the fraction of receivers that can decode an equivalent of a single packet from a batch of K packets.

We numerically evaluate r'[s] as a function of s for different values of K as shown in Fig. 9.8 for the case where $c = 0.5$. It can be seen from the graph that for $K = 30$, a large fraction of users are served within a short duration and then the source takes a relatively longer time to serve the remaining small fraction of users towards the end of the transmission of the current block of K packets. On increasing K to 60 and then to 120, the transition becomes increasingly sharper indicating that the source serves a larger fraction of users in a shorter duration and takes less time to serve a smaller fraction of users towards the end of the transmission. Ideally, we would like all the users to complete decoding together for an increase in throughput. This can be achieved by increasing K indefinitely as observed from Fig. 9.8. However, this causes the decoding delay to increase indefinitely as well. Hence, it is important to understand the throughput-delay trade-off as K scales as a function of N.

This important problem is studied in a recent work [30], where the throughput and delay performance of RNC is studied through an appropriate Gaussian approximation of the key system parameters, leading to tractable and tight bounds on the approximate behavior. In particular, the

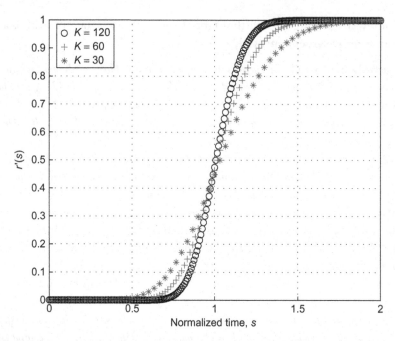

Figure 9.8 Fraction of receivers that have successfully decoded a single packet in a block of K packets in s time slots, $r'[s]$ as a function of s for $c = 1/2$.

study reveals that the coding window size (and hence decoding delay) of $K = \Theta(\ln(N))$ represents a phase transition rate below which the throughput converges to zero, and above which it converges to the broadcast capacity (i.e., the maximum rate at which data can be broadcast to all N users over the erasure channels) of c. Before providing this result, we describe the suggested Gaussian approximation.

Definition 3. *We define approximate decoding delay* $\tilde{Z}(N, K)$ *and approximate mean throughput* $\tilde{\tau}^{RNC}(N, K)$ *as follows:*

$$\tilde{Z}^{RNC}(N, K) = \frac{K}{c} + \frac{\sqrt{K(1-c)}}{c} \max_{1 \leq i \leq N} \tilde{\chi}_i, \tag{3}$$

$$\tilde{\tau}^{RNC}(N, K) = \frac{K}{\mathbb{E}[\tilde{Z}^{RNC}(N, K)]}, \tag{4}$$

where $\tilde{\chi}_i, i = 1, 2, \ldots, N$ *are independent standard normal random variables.*

Theorem 5. *The approximate mean throughput $\tilde{\tau}^{RNC}(N, K)$ shows the following behavior:*
1. *When $K = o(\ln(N))$, then $\tilde{\tau}^{RNC}(N, K)$ goes to zero as $N \to \infty$.*
2. *When $K = \Theta(\ln(N))$, then $\tilde{\tau}^{RNC}(N, K)$ approaches the broadcast capacity c as $N \to \infty$.*
3. *When $K = \omega(\ln(N))$, then $\tilde{\tau}^{RNC}(N, K)$ approaches c as $N \to \infty$.*

The predictions of this theorem are confirmed in Figs. 9.9–9.11 for the i.i.d. erasure channels with $c = 0.9$. In particular, Fig. 9.9 confirms the phase transition nature of the throughput behavior by studying various scaling types of K with respect to N. As suggested by the Gaussian approximation, $K = \Theta(\log(N))$ represents a sharp phase transition scaling above which maximum possible throughput of $c = 0.9$ is asymptotically achieved, and below which the throughput asymptotically vanishes.

Figures 9.10 and 9.11 study the proximity of the actual system performance to the Gaussian approximation provided above, and check the

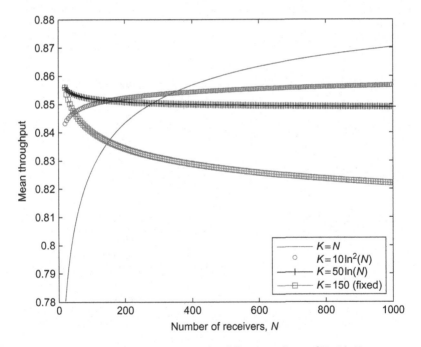

Figure 9.9 Mean throughput behavior under different scalings of K with N.

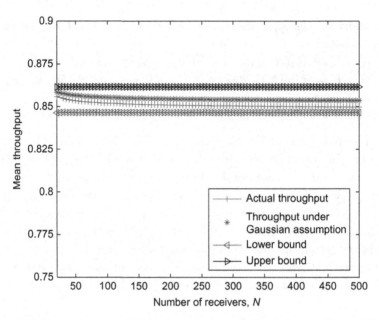

Figure 9.10 Comparing actual throughput to upper and lower bounds for $c = 0.9$ and $K = 50\ln(N)$.

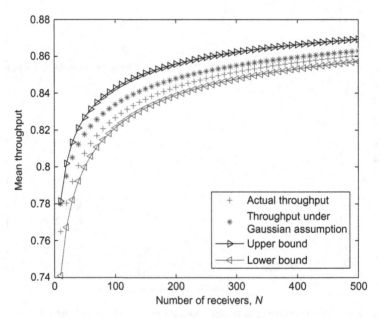

Figure 9.11 Comparing actual throughput to upper and lower bounds for $c = 0.9$ and $K = N$.

validity of the bounds of the approximate performance to the actual performance. These numerical studies not only support the validity of the approximate results of the theorem, but also reveal the sharpness of the approximate upper and lower bounds. In a more recent work [31], tight upper and lower bounds are derived on the actual system performance to show that the asymptotic results of Theorem 5 based on the above Gaussian approximation are in fact accurate for the real system.

3. CODING IN LARGE-SCALE MOBILE *AD HOC* NETWORKS

Another important application of network coding is in mobile *ad hoc* networks (MANETs). MANETs represent one of the most innovative emerging networking technologies, with broad potential applications in personal area networks, emergency and rescue operations, and military battlefield applications, etc. For example, the ZebraNet [36] is a MANET used to monitor and study animal migrations and inter-species interactions, where each zebra is equipped with an wireless antenna and pairwise communication is used to transmit data when two zebras are close to each other. Another example is the mobile–phone mesh network proposed by TerraNet AB (a Swedish company) [29], where the participating mobile phones form a mesh network and can talk to each other without using the cell infrastructure.

Because of their important applications in various areas, understanding the fundamental limits of MANETs has received significant attention in the past few years. A widely studied model of MANET is a mobile network with n mobiles deployed in a unit torus. The MANET contains n flow sessions. Each mobile is a source of a session and a destination of another session. Further, the mobility patterns of the mobiles are independently and identically distributed.

The throughput scaling of this model is first characterized in [13], where the authors showed that $\Theta(1)$ throughput per session is achievable. We note that in a static *ad hoc* wireless network with n nodes and n flow sessions, the throughput per session is shown to be at most $\Theta(1/\sqrt{n})$ in the seminal work [14] of Gupta and Kumar. The key idea in [13] is to exploit mobile relays to carry packets close to their destinations and then deliver the packets. This approach significantly increases the throughput, but also leads to large delays (a $\Theta(n \log n)$ delay under a random walk model [8]). Since then, a number of papers have investigated the trade-offs between throughput and delay [4, 6–9, 13, 20–23, 28, 32]. The widely used approach in these papers is to relay a packet to multiple mobiles so that the packet can be delivered when one of the mobiles gets close to the destination. This approach clearly reduces the per-packet delivery time (delay) but results in a smaller throughput due to duplication.

In a recent work [34], the idea of using *linear network coding* in MANETs is developed. The algorithms result in orders of magnitude of performance improvement compared to those algorithms without coding. The following example presented in [34] illustrates the key intuition of using coding in MANETs.

3.1. An Example: Delay-Throughput Trade-off under an i.i.d. Mobility Model

Consider a mobile network with n mobiles deployed in a unit torus. Assume that node i sends information to node $(i + 1)$ mod (n), so there are n data flows in the network. Node i and node $(i + 1)$ mod (n) are called a source-destination. The mobiles move according to an i.i.d. mobility model.

Definition 4 (I.I.D. Mobility Model). *The n wireless mobile nodes are uniformly, randomly positioned on a unit torus. The node positions are independent of each other, and independent from time slot to time slot. Therefore, under the i.i.d. mobility model, the positions of the mobiles are totally reshuffled at each time slot.*

Further, assume the wireless interference can be modeled using the protocol model introduced in [1].

Definition 5 (Protocol Model [1]). *Let α_i denote the transmission radius of node i, then a transmission from node i to node j is successful under the protocol model if, and only if, the following two conditions hold: (i) the distance between nodes i and j is less than α_i, and; (ii) if mobile k is transmitting to mobile h at the same time, then the distance between node k and node j is at least $(1 + \Delta)\alpha_k$, where $\Delta > 0$ defines a guard zone around the transmission.*

In MANETs, there are two fundamental constraints that limit their performance:
- *wireless interference* which limits the number of simultaneous transmissions during each time slot; and
- *mobility* which determines the distance a packet needs to travel to get to the destination.

At first glance, these two constraints seem quite different, which makes the capacity characterization of MANETs a very challenging problem. The authors of [34] overcame this hurdle by using a virtual channel framework to unify both constraints. The virtual channel framework reveals the critical role of linear network coding in MANETs.

To simplify our analysis, assume that the packets are transmitted using the two-hop transmission algorithm introduced in [13].

Definition 6 (Two-hop transmission algorithm). *The two-hop transmission algorithm works as follows:*
- *At the first hop, a packet is transmitted from its source to a relay (or several relays) around the source; and*
- *at the second hop, a packet is transmitted from a relay to its destination when the relay is close enough to the packet's destination.*

This two-hop transmission algorithm does not allow multihop relay, but the algorithm nevertheless captures three critical phases of a successful delivery (see Fig. 9.12):
- Phase-I: the packet is transmitted from the source to one or multiple relay nodes, and the transmission radii are assumed to be L_1;
- Phase-II: a relay moves to the neighborhood of the destination of the packet; and

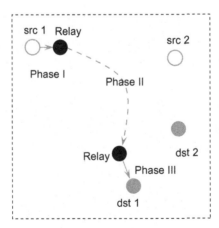

Figure 9.12 The three phases of a typical delivery under the two-hop transmission scheme, in which the dotted line indicates the movement of the relay node over one or multiple time slots.

- Phase-III: the relay sends the packet to the destination, and the transmission radii in this phase are assumed to be L_2.

In [34], the authors model each phase as a virtual communication channel:

- **Reliable broadcasting channel:** Without loss of generality, it can be assumed that $\Delta = 2$. Under the protocol model, the network can support at most $\frac{1}{\pi(L_1)^2}$ simultaneous transmissions. On average, a mobile gets:

$$P_1 = \frac{1}{\pi(L_1)^2 n}$$

fraction of time to broadcast its packets to relays, and a packet is relayed to $\pi(L_1)^2 n$ mobiles on average. The capacity of a virtual channel is:

$$\frac{W}{\pi(L_1)^2 n},$$

where W is the maximum number of packets that can be transmitted from one mobile to another in one time slot.

- **Unreliable relay channel (erasure channel):** A relay can transmit a packet to its destination if the relay is within a distance of L_2 from the destination. If this event occurs, we say the relayed packet is deliverable. Under the i.i.d. mobility model, a packet becomes deliverable with probability:

$$\pi(L_2)^2.$$

Recall that each packet is relayed to $\pi(L_1)^2 n$ mobiles on average. The probability that none of them is deliverable in D consecutive time slots is:

$$\left(1 - \pi(L_2)^2\right)^{\pi(L_1)^2 nD}.$$

Therefore, the relay phase can be modeled as an erasure channel with erasure probability:

$$P_{\text{erasure}} = \left(1 - \pi(L_2)^2\right)^{\pi(L_1)^2 nD}.$$

- **Reliable receiving channel:** Consider the transmissions from relays to destinations. Under the protocol model, the number of simultaneous deliveries is no more than:

$$\frac{1}{\pi(L_2)^2}.$$

Each destination gets:

$$\frac{1}{\pi(L_2)^2 n}$$

fraction of time to receive packets. So the capacity of this receiving channel is:

$$\frac{W}{\pi(L_2)^2 n}.$$

Based on the virtual channel argument above, the MANET can be represented as concatenation of the three virtual channels as shown in Fig. 9.13. Based on this heuristic argument, it can be computed that the maximum throughput per session is:

$$\lambda = \max_{L_1,L_2} \min \left\{ \left(1 - \left(1 - \pi(L_2)^2\right)^{\pi D(L_1)^2 n}\right) \frac{W}{\pi n(L_1)^2}, \frac{W}{n\pi(L_2)^2} \right\}$$

$$= \Theta\left(\sqrt{\frac{D}{n}}\right),$$

where the transmission radii L_1 and L_2 that solve the maximization problem are $L_1^* = \Theta\left(\frac{1}{\sqrt[2]{n}}\right)$ and $L_2^* = \Theta\left(\frac{1}{\sqrt[4]{Dn}}\right)$. Clearly to achieve this throughput, *a coding scheme achieving the capacity of the erasure channel* is needed. Since the erasure probability is determined by L and D, which are different under different delay constraints, *linear network coding* becomes the first choice.

We would like to emphasize that the virtual channel not only provides a unified view for understanding the limits of MANETs, but also reveals

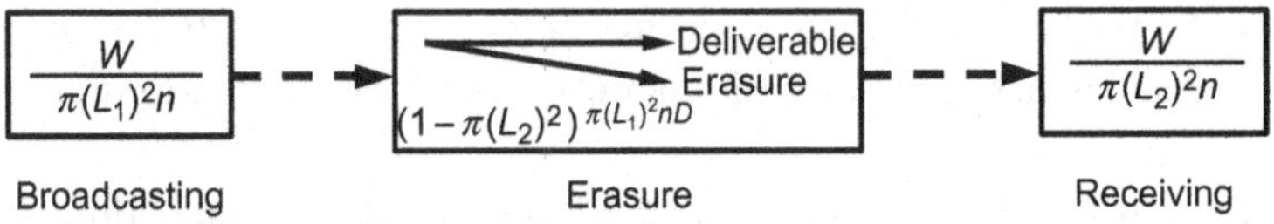

Figure 9.13 The virtual channel representation.

the importance of linear network coding in MANETs. Based on this heuristic algorithm, the authors in [34] proposed the following joint coding-scheduling algorithm and proved that the algorithm achieves the optimal delay-throughput trade-off in an order sense.

Joint Coding-Scheduling Algorithm [34]

We divide the unit square into square cells with each side of length equal to $1/\sqrt[4]{nD}$. The transmission radius of each node is chosen to be $\sqrt{2}/\sqrt[4]{nD}$, so that any two nodes within a cell can communicate with each other. This means that, given the interference constraint, two nodes in a cell can communicate if all nodes in cells within a fixed distance from the given cell stay silent. Each time slot is further divided into 9 minislots and each cell is guaranteed to be active in at least one minislot within each time slot. The reason we use nine minislots is that if a node in a cell is active, then no other nodes in any of its neighboring eight cells can be active, but nodes outside this neighborhood can be active. Further, we denote the packet size to be $W/18$ so that two packets can be transmitted in each minislot. We group every $6D$ time slots into a supertime slot. At each supertime slot, the nodes transmit packets as follows.

1. **Random linear encoding:** Each source takes $6D/(25M)$ data packets, and uses random linear coding to generate D/M coded packets, where $M = \sqrt{n/D}$.
2. **Broadcasting:** This step consists of D time slots. At each time slot, the nodes execute the following tasks:
   (i) A cell is said to be a good cell at time t if the number of nodes in the cell is between $9M/10$ and $11M/10$. In each good cell, one node is randomly selected. If the selected node has not already transmitted all of its D/M coded packets, then it broadcasts a coded packet that was not previously transmitted to $9M/10$ other nodes in the cell during the minislot allocated to that cell. Recall that our choice of packet size allows one node in every good cell to transmit during every time slot.
   (ii) All nodes check the duplicate packets they have. If two or more packets have the same destination, select one at random and drop the others.
3. **Receiving:** This step consists of $5D$ time slots. At each time slot, if a cell contains no more than two deliverable packets, the deliverable packets are delivered to their destinations using one-hop transmissions

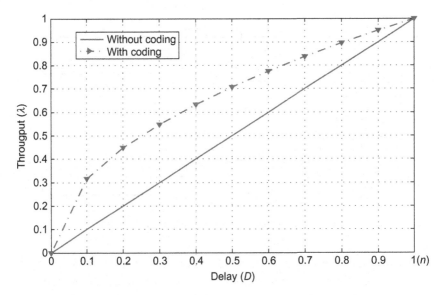

Figure 9.14 Delay-throughput trade-offs with coding and without coding.

during the minislot allocated to that cell; otherwise, no node in the cell attempts to transmit. At the end of this step, all undelivered packets are dropped. The destinations decode the received coded packets.

Theorem 6. *Under the two-dimensional i.i.d. mobility model, the throughput per source-destination pair is $\lambda = O\left(\sqrt{D/n}\right)$ given delay constraint 6D. When D is both $\omega(\sqrt[3]{n})$ and $o(n)$, this throughput can be achieved using the joint coding-scheduling algorithm.*

We note that without using linear network coding, from the best of our knowledge, the best delay-throughput trade-off is $\lambda = \Theta(D/n)$, which was established in [23]. Figure 9.14 is a comparison between D/n and $\sqrt{D/n}$ when D increases from 0 to n. The figure shows a significant gain of using coding.

3.2. Extension to Multicast Traffic Flows

This framework can be extended to the multicast case. In [37], the authors consider a network with n_s multicast sessions where each multicast session consists of single source and p destinations. Similar to the unicast scenario, the authors constructed a virtual channel system as shown in Fig. 9.15.

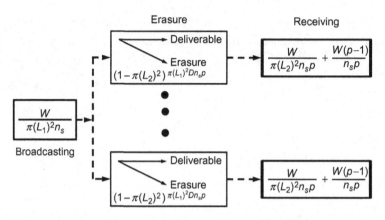

Figure 9.15 The virtual channel representation of a multicast session.

Note that this virtual system is similar to the broadcast network presented in Section 2. Therefore, linear network coding can again be exploited to significantly improve the performance of this multicast MANET.

Note that in the multicast case, all destinations in the same session are interested in the same information. When a packet is delivered to one of its destinations, other destinations within the transmission range can receive the packet, which results in multiple deliveries. This is different from the unicast case. The authors named one of the deliveries as *target delivery*, and the rest as *free-ride deliveries*. Under the protocol model, all exclusion regions associated with the targeted deliveries should be disjoint from each other. With a common transmission radius L_2, assume that a successful target delivery associates an exclusion region with area $\pi(L_2)^2$. So the number of simultaneous target deliveries is no more than:

$$\frac{W}{\pi(L_2)^2}.$$

Furthermore, along with each target delivery, there are:

$$(p-1)\pi(L_2)^2$$

free-ride deliveries on average. Thus, the average number of deliveries per time slot is:

$$\frac{W(1+(p-1)\pi(L_2)^2)}{\pi(L_2)^2}.$$

Recall that a source packet needs to be delivered to all p destinations, so the throughput per multicast session is:

$$W\frac{1 + (p-1)\pi(L_2)^2}{n_s p\pi(L_2)^2} = \frac{W}{n_s p\pi(L_2)^2} + \frac{W(p-1)}{n_s p}$$

packets per time slot. Based on the virtual channel representation, it can be heuristically computed that:

$$\lambda = \max_{L_1,L_2} \min \left\{ \left(1 - \left(1 - \pi(L_2)^2\right)^{\pi D(L_1)^2 n}\right) \frac{W}{\pi(L_1)^2 n_s}, \right.$$
$$\left. \frac{W}{n_s p\pi(L_2)^2} + \frac{W(p-1)}{n_s p} \right\}$$
$$= \Theta\left(\sqrt{\frac{D}{n_s}}\right),$$

where the transmission radii L_1 and L_2 that solve the maximization problem are $L_1^* = \Theta\left(\frac{1}{\sqrt[2]{n_s}}\right)$ and $L_2^* = \Theta\left(\frac{1}{\sqrt[4]{p^2 D n_s}}\right)$.

Theorem 7. *Under the two-dimensional i.i.d. mobility model, the throughput per multicast session is $\lambda = O\left(\sqrt{D/n_s}\right)$ given delay constraint D. When D is both $\omega(\sqrt[3]{n})$ and $o(n)$, this throughput can be achieved using a joint coding-scheduling algorithm presented in [37].*

3.3. Summary of Existing Results

Exploiting coding to improve the efficiency of MANETs has been exploited in several other papers. The results have been summarized below:

1. Two-dimensional i.i.d. mobility models with unicast traffic:
   (i) Under a fast mobility assumption, it is shown in [34] that the maximum throughput per source-destination pair is $O\left(\sqrt{D/n}\right)$ under a delay constraint D. In the same paper, a joint coding-scheduling algorithm is presented to achieve the maximum throughput when D is both $\omega\left(\sqrt[3]{n}\right)$ and $o(n)$. In [35], it has shown that $\Theta(1)$ is achievable with $D = \Theta(n)$, and $\Theta(1/n)$ is achievable with $D = \Theta(\log n)$.

(ii) Under a slow mobility assumption, it is shown in [34] that the maximum throughput per source-destination pair is $O\left(\sqrt[3]{D/n}\right)$ given a delay constraint D. A joint coding-scheduling algorithm to achieve the maximum throughput when D is both $\omega(1)$ and $o(n)$ is proposed in [34].

2. Two-dimensional hybrid random walk models with unicast traffic:

(i) Under the fast mobility assumption, it is shown in [34] that the maximum throughput per source-destination pair is $O(\sqrt{D/n})$ when $S = o(1)$ and $D = \omega(|\log S|/S^2)$, where S is the step-size of the random walk. A joint coding-scheduling algorithm is proposed in [34] to achieve the maximum throughput when $S = o(1)$ and D is both $\omega(\max\{(\log^2 n)|\log S|/S^6, \sqrt[3]{n}\log n\})$ and $o(n/\log^2 n)$. When $S = 1/\sqrt{n}$, in [18, 35], it has been shown that $\Theta(1)$ is achievable with $D = \Theta(n)$; and in [35], throughput $\Theta(1/n)$ is proved to be achievable given $D = \Theta(\log n)$.

(ii) Under the slow mobility assumption, it is shown in [34] that the maximum throughput per source-destination pair is $O(\sqrt[3]{D/n})$ when $S = o(1)$ and $D = \omega(|\log S|/S^2)$, and a joint coding-scheduling algorithm is proposed to achieve the maximum throughput when $S = o(1)$ and D is both $\omega((\log^2 n)|\log S|/S^6)$ and $o(n/\log^2 n)$.

3. One-dimensional i.i.d. mobility models with unicast traffic:

(i) Under the fast mobility assumption, it is shown in [34] that the maximum throughput per source-destination pair is $O\left(\sqrt[3]{D^2/n}\right)$ given a delay constraint D. A joint coding-scheduling algorithm is proposed in [34] to achieve the maximum throughput when D is both $\omega(\sqrt[5]{n})$ and $o\left(\sqrt{n}/\sqrt[3/2]{\log n}\right)$.

(ii) Under the slow mobility assumption, it is shown in [34] that the maximum throughput per source-destination pair is $O\left(\sqrt[4]{D^2/n}\right)$. A joint coding-scheduling algorithm is proposed in [34] to achieve the maximum throughput when D is $o\left(\sqrt{n}/\log^2 n\right)$.

4. One-dimensional hybrid random walk models with unicast traffic:

(i) Under the fast mobility assumption, it is shown in [34] that the maximum throughput per source-destination pair is $O(\sqrt[3]{D^2/n})$ when $S = o(1)$ and $D = \omega(1/S^2)$, and a joint coding-scheduling algorithm is proposed in [34] to achieve the maximum

Table 9.2 Delay-throughput trade-offs with coding and without coding in MANETs

Mobility model	With coding	Without coding
2D-i.i.d. (fast&unicast)	$\Theta\left(\sqrt{\frac{D}{n}}\right)$ [34, 35]	$\Theta\left(\frac{n}{D}\right)$ [23]
2D-i.i.d. (slow&unicast)	$\Theta\left(\sqrt[3]{\frac{D}{n}}\right)$ [34]	$\Theta\left(\sqrt[3]{\frac{D}{n\log^{9/2}n}}\right)$ [21]
Random-walk (slow&unicast)	$\lambda = \Theta(1)$ when $D = \Theta(n)$ [18, 35]	$\lambda = \Theta(1)$ when $D = \Theta(n\log n)$ [8]
2D-i.i.d. (fast&multicast)	$\Theta\left(\sqrt[3]{\frac{D}{n_s}}\right)$ [37]	$\lambda = \dfrac{1}{p\sqrt{n_s p \log p}}$ when $D = \Theta(\sqrt{n_s p \log p})$ [16]

throughput when $S = o(1)$ and D is both $\omega(\max\{(\log^2 n) \mid \log S \mid / S^4, \sqrt[5]{n}\log n\})$ and $o\left(\sqrt{n}/\sqrt[3/2]{\log n}\right)$.

(ii) Under the slow mobility assumption, it is shown in [34] that the maximum throughput per source-destination pair is $O\left(\sqrt[4]{D^2/n}\right)$ when $S = o(1)$ and $D = \omega(1/S^2)$, and a joint coding-scheduling algorithm is proposed in [34] to achieve the maximum throughput when $S = o(1)$ and D is both $\omega\left((\log^2 n)\mid\log S\mid/S^4\right)$ and $o\left(\sqrt{n}/\log^2 n\right)$.

5. Two-dimensional i.i.d. mobility models with multicast traffic:

(i) Given a delay constraint D, in [37], the throughput per multicast session is proved to be $O\left(\min\left\{1, (\log p)(\log (n_s p))\sqrt{\frac{D}{n_s}}\right\}\right)$. In the same paper, a joint coding-scheduling algorithm achieving a throughput of $\Theta\left(\min\left\{1, \sqrt{\frac{D}{n_s}}\right\}\right)$ is proposed.

Table 9.2 compares the delay-throughput trade-offs that can be achieved with coding with those without coding. The conditions under which these trade-offs are achievable are omitted here, the reader can find the details in the corresponding references.

4. CONCLUSION

In this chapter, we have summarized some of the interesting findings on the performance of network coding in unreliable wireless networks under several scaling regimes, including the network size, the coding window size, the number of flows in the network, and the application delay constraints.

These results not only help quantify the performance of network coding for comparison to traditional scheduling and routing policies, but also identify cases in which significant throughput, delay, or economic gains are achievable even with simple randomized network coding strategies. We end this chapter by noting that this is still a very active area of research after its decade-long lifetime, and that our overview only contains a subset of the existing and ongoing works in this exciting field. We believe that many interesting results are due in this area, and hope that the current chapter helps motivate current and future researchers to pursue them.

REFERENCES

[1] A. Agarwal and P. R. Kumar. Improved capacity bounds for wireless networks. *Wireless Communications and Mobile Computing*, 4: 251–261, 2004.

[2] R. Ahlswede, Ning Cai, S. R. Li, and R. W. Yeung. Network information flow. *IEEE Transactions on Information Theory*, 46: 1204–1216, July 2000.

[3] E. Ahmed, A. Eryilmaz, A. Ozdaglar, and M. Médard. Economic aspects of network coding. In *Forty-fourth Annual Allerton Conference on Communication, Control and Computing*, Monticello, IL, September 2006.

[4] N. Bansal and Z. Liu. Capacity, delay and mobility in wireless *ad hoc* networks. In *Proc. IEEE Infocom.*, San Francisco, CA, 2003.

[5] H. A. David. *Order Statistics*. Wiley Series in Probability and Statistics, 1981.

[6] S. N. Diggavi, M. Grossglauser, and D. Tse. Even one-dimensional mobility increases *ad hoc* wireless capacity. In *Proc. IEEE Int. Symp. Information Theory (ISIT)*, page 352, July 2002.

[7] A. El-Gamal, J. Mammen, B. Prabhakar, and D. Shah. Throughput-delay trade-off in wireless networks. In *Proc. IEEE Infocom.*, pages 475–485, 2004.

[8] A. El-Gamal, J. Mammen, B. Prabhakar, and D. Shah. Optimal throughput-delay scaling in wireless networks – part I: The fluid model. *IEEE Trans. Inform. Theory*, 52(6): 2568–2592, June 2006.

[9] A. El-Gamal, J. Mammen, B. Prabhakar, and D. Shah. Optimal throughput-delay scaling in wireless networks – part II: Constant-size packets. *IEEE Trans. Inform. Theory*, 52(11): 5111–5116, November 2006.

[10] A. Eryilmaz, A. Ozdaglar, and M. Médard. On delay performance gains from network coding. In *Proceedings of Conference on Information Sciences and Systems (CISS)*, Princeton, NJ, March 2006.

[11] A. Eryilmaz, A. Ozdaglar, M. Médard, and E. Ahmed. On the delay and throughput gains of coding in unreliable networks. *IEEE Transactions on Information Theory*, 54: 5511–5524, December 2008.

[12] M. Ghaderi, D. F. Towsley, and J. F. Kurose. Reliability gain of network coding in lossy wireless networks. In *INFOCOM*, pages 2171–2179, 2008.

[13] M. Grossglauser and D. Tse. Mobility increases the capacity of *ad hoc* wireless networks. In *Proc. IEEE Infocom.*, volume 3, pages 1360–1369, April 2001.

[14] P. Gupta and P. Kumar. The capacity of wireless networks. *IEEE Trans. Inform. Theory*, 46(2): 388–404, 2000.

[15] T. Ho and D. Lun. *Network Coding: An Introduction*. Cambridge University Press, 2008.

[16] C. Hu, Xinbing Wang, and F. Wu. Motioncast: On the capacity and delay trade-offs. In *ACM MobiHoc 09*, New Orleans, May 2009.

[17] R. Koetter and M. Médard. Beyond routing: An algebraic approach to network coding. *IEEE Transactions on Information Theory*, 11: 782–795, October 2003.

[18] Z. Kong, E.M. Yeh, and E. Soljanin. Coding improves the throughput-delay trade-off in mobile wireless networks. In *Proc. IEEE Int. Symp. Information Theory (ISIT)*, pages 1784–1788, 2009.

[19] S.-Y. R. Li, R. W. Yeung, and Ning Cai. Linear network coding. *IEEE Transactions on Information Theory*, 49: 371–381, February 2003.

[20] X. Lin, G. Sharma, R. R. Mazumdar, and N. B. Shroff. Degenerate delay-capacity trade-offs in *ad hoc* networks with Brownian mobility. *Joint Special Issue of IEEE Transactions on Information Theory and IEEE/ACM Transactions on Networking on Networking and Information Theory*, 52(6): 2777–2784, June 2006.

[21] X. Lin and N. Shroff. Towards achieving the maximum capacity in large mobile wireless networks. *J. Commun. and Networks*, 4: 352–361, 2004.

[22] J. Mammen and D. Shah. Throughput and delay in random wireless networks with restricted mobility. *IEEE Trans. Inform. Theory*, 53(3): 1108–1116, March 2007.

[23] M.J. Neely and E. Modiano. Capacity and delay trade-offs for *ad hoc* mobile networks. *IEEE Trans. Inform. Theory*, 51(6): 1917–1937, 2005.

[24] D. Nguyen, T. Nguyen, and B. Bose. Wireless broadcasting using network coding. In *Proceeding of NetCod*, 2007.

[25] S. I. Resnick. *Extreme Values, Regular Variation, and Point Processes*. Springer, 1987.

[26] S. M. Ross. *Stochastic Processes*. John Wiley & Sons, 1995.

[27] C. E. Shannon. A mathematical theory of commmunication. *Bell Systems Technical Journal*, 27: 379–423, 623–656, July–October 1948.

[28] G. Sharma, R. Mazumdar, and N. Shroff. Delay and capacity trade-offs in mobile *ad hoc* networks: A global perspective. In *Proc. IEEE Infocom.*, pages 1–12, April 2006.

[29] TerraNet. http://terranet.se/.

[30] S. B. Tirumala, A. Eryilmaz, and N. Shroff. Throughput-delay analysis of random linear network coding for wireless broadcasting. In *IEEE International Symposium on Network Coding (NetCod)*, 2010.

[31] S. B. Tirumala, A. Eryilmaz, and N. Shroff. Throughput-delay analysis of random linear network coding for wireless broadcasting, 2011. Technical Report, submitted to IEEE Transactions on Information Theory.

[32] S. Toumpis and A. J. Goldsmith. Large wireless networks under fading, mobility, and delay constraints. In *Proc. IEEE Infocom.*, volume 1, pages 619–627, 2004.

[33] W. Xiao and D. Starobinski. Extreme value fec for reliable broadcasting in wireless networks. In *INFOCOM*, 2009.

[34] L. Ying, S. Yang, and R. Srikant. Optimal delay-throughput trade-offs in mobile *ad hoc* networks. *IEEE Trans. Inform. Theory*, 9(54):4119–4143, September 2008.

[35] C. Zhang, X. Zhu, and Y. Fang. On the improvement of scaling laws for large-scale MANETs with network coding. *IEEE Journal on Selected Areas in Communications*, 27(5): 662–672, 2009.

[36] P. Zhang, C. M. Sadler, S. A. Lyon, and M. Martonosi. Hardware design experiences in zebranet. In *Proc. the 2nd International Conference on Embedded Networked Sensor Systems*, pages 227–238, 2004.

[37] S. Zhou and L. Ying. On delay constrained multicast capacity of large-scale mobile *ad hoc* networks. In *Proc. IEEE Infocom. Mini-Conference*, San Diego, CA, 2010.

CHAPTER *10*

Network Coding in Disruption Tolerant Networks

Xiaolan Zhang[1], Giovanni Neglia[2], and Jim Kurose[3]
[1]Department of Computer & Information Sciences, Fordham University, New York, USA;
[2]INRIA, Sophia Antipolis, France;
[3]Department of Computer Science, University of Massachusetts, Amherst, MA, USA

Contents

Network Coding. DOI: 10.1016/B978-0-12-380918-6.00010-X
Copyright © 2012 Elsevier Inc. All rights reserved.

Abstract

Disruption Tolerant Network (DTN) scenarios often arise from mobile wireless networks, where due to limited transmission power, fast node mobility, sparse node density, and frequent equipment failures, there is often no contemporaneous path from the source to the destination node(s). We review research works that studied the benefits of applying Random Linear Coding (RLC) to DTNs in this chapter. We first review traditional non-coding routing schemes for broadcast and unicast applications in DTNs, and the basic operations of RLC. After introducing the design space of DTN routing, we then focus on the performance evaluation of RLC based routing schemes for broadcast communication and unicast communication. For both communications, the RLC based schemes improve the trade-off between energy expenditure and delivery performance. In addition, research efforts in the performance modeling of RLC schemes, priority coding protocol, RLC based secure unicast scheme are also summarized. The chapter ends with a discussion about open issues on the application of network coding to DTNs.

Keywords: Network coding, random linear coding, disruption tolerant network, performance trade-off, broadcast communication, unicast communication.

1. INTRODUCTION

In recent years, wireless communication technologies have been increasingly deployed in challenging environments where there is no communication infrastructure, as evidenced by the many efforts in building and deploying wireless sensor networks for wildlife tracking [22, 43], underwater sensor networks [38, 41], disaster relief team networks, networks for remote areas or for rural areas in developing countries [8, 10, 51], vehicular networks [4, 20], and Pocket Switched Networks [19]. Without infrastructure support, such networks rely solely on peer-to-peer connectivity among wireless radios to support data communication. Owing to the limits of transmission power, fast node mobility, sparse node density, and frequent equipment failures, many such networks have only intermittent connectivity, and experience frequent disconnection of nodes. *Disruption Tolerant Network* (DTN, or *Delay Tolerant Network*), refers to such a network where there is often no contemporaneous path from the source node to the destination node. End-to-end communication in DTNs adopts a so-called "store-carry-forward" paradigm—a node receiving a packet buffers and carries the packet as it moves, passing the packet on to new nodes that it encounters. When the destination node meets a node that carries the packet, the packet is delivered to the destination.

In addition to intermittent connectivity and dynamic topologies, routing in DTNs faces additional challenges due to the severe resource constraints: for the small mobile nodes carried by animals or humans, buffer space, transmission bandwidth, and power are very limited; for mobile nodes in vehicle based networks, even though buffer space or power are usually not severely constrained, transmission bandwidth is still a scarce resource.

There has been a substantial amount of research on the benefits of network coding for wireless networks. For multicast applications in static wired or wireless networks, Lun et al. [35] and Wu et al. [52] showed that for the problem of minimum-energy multicast, the use of network coding simplifies the problem (from an NP-complete problem to a linear optimization problem solvable in polynomial time). For broadcast applications in mobile and static wireless networks, Widmer et al. [49, 50] proposed a Random Linear Coding (RLC) [17, 18] based scheme for energy efficient broadcast. For unicast applications in static wireless networks, several works [24, 25, 34, 53] have shown that network coding schemes can provide throughput gain by leveraging the broadcast nature of the wireless channel.

Because of the distinct characteristics of DTNs, some of the benefits of network coding identified above for general wireless networks do not hold for DTNs. First, due to the dynamically changing topology of DTNs, the static network model adopted in [35, 52] and the results obtained therein about coding benefits for multicast applications are not directly applicable to DTNs. Secondly, DTNs have sparse node density, with each node usually having at most one neighbor at any instance of time, therefore the previously discovered benefit of network coding in increasing network throughput (by leveraging the broadcast nature of wireless transmission) is negligible for DTNs. On the other hand, there are new opportunities for network coding in DTNs. The rapidly changing topology and the lack of infrastructure require DTN routing schemes to be *distributed*; moreover the limited connectivity and bandwidth require DTN routing schemes to be *localized* (i.e., with only limited knowledge about the local neighborhood) too. Network coding, in addition to its benefits in increasing throughput and cost saving, has been shown to facilitate the design of efficient distributed routing schemes [36].

Existing research on the application of network coding to DTNs has focused on applying Random Linear Coding (RLC) to broadcast and unicast communication. In what follows we will use the expression *RLC scheme* to denote a DTN routing scheme that employs RLC, and use the expression *non-coding scheme* to denote a traditional routing scheme. For broadcast applications where all nodes are interested in receiving all packets, Widmer *et al.* [49, 50] demonstrated the benefit of RLC schemes in improving energy efficiency. For unicast transmissions, [55, 57] demonstrated that RLC schemes can achieve faster propagation of a block of unicast packets. Furthermore, RLC schemes, when combined with binary *spray-and-wait* schemes to control the number of transmissions made in the network, improve the trade-off between delivery delay and overhead [31, 55, 57]. References [30, 31] proposed a modeling study of RLC schemes. Finally, for DTNs with Byzantine adversaries, [42] proposed a network coding based routing scheme that supports secure data communication.

The rest of the chapter is structured as follows. In Section 2, we present basic background on DTNs, non–coding based broadcast and unicast routing schemes, and Random Linear Coding. In Section 3, we introduce a taxonomy of DTN routing schemes by discussing the various design aspects of DTN routing schemes. Section 4 and Section 5, respectively, present research contributions on the benefit of network coding for broadcast application and unicast application in DTNs. Section 6 discusses open issues about the application of network coding to DTNs. Finally, Section 7 summarizes the chapter.

2. BACKGROUND ON DISRUPTION TOLERANT NETWORKS AND RANDOM LINEAR CODING

In this section, we first introduce the network model, then review non-coding based routing schemes that have been proposed for broadcast and unicast communication in DTNs, and finally provide an introduction to the basic operations of RLC.

2.1. Network Model

Consider a network consisting of $N + 1$ mobile nodes moving independently in a closed area according to some common mobility model such as random waypoint or random direction model [5]. Each node is equipped with a wireless radio with a common transmission range so that when two nodes come within transmission range of each other (they *meet*), they can exchange packets with each other. The *meeting time* of these two nodes is the time duration of this transmission opportunity, while the *inter-meeting time* is the duration of the time interval between two consecutive meetings, i.e., from the time instant when the two nodes go out of transmission range of each other to the time instant when they can again communicate. It has been shown in [12] that under random waypoint and random direction models, the inter-meeting time follows approximately an exponential distribution when node velocity is relatively large compared to the region size and the transmission range is relatively small. Because of the tractability of the exponential inter-meeting time mobility model, it has been widely adopted (see e.g., [31, 55, 57] and references therein.).

We refer to the list of node-to-node contacts of a DTN during a certain time duration as *a DTN contact trace*, and use the so-called *temporal network* model that was originally proposed by Kempe *et al.* [26] to represent a DTN contact trace. The temporal network is a multi-graph $G = (V, E)$ in which V denotes the nodes in the network, and E denotes the set of edges, with each edge representing a node-to-node contact. Each edge $e \in E$ is labeled with a pair, $(t(e), bw(e))$, where $t(e)$ specifies the time at which the two endpoint nodes are able to communicate, and $bw(e)$ specifies the bandwidth constraint of the contact, i.e., the number of packets that can be exchanged over the contact. The edges can be *directed*, if independent wireless channels are used for transmissions in the two directions, or *undirected*, if the same wireless channel is used for transmission in both directions, and the total amount of capacity can be arbitrary divided between them. For example, Fig. 10.1 illustrates the temporal network model with directed edges for a contact trace of a DTN with four nodes during time interval $[0, 24]$.

Existing works study both broadcast and unicast communications in DTNs. Broadcast communication delivers each message to all nodes in the network, as in the case of content distribution service, and routing message propagation; while under unicast communication, each message is destined

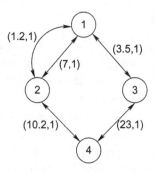

Figure 10.1 Temporal network representing the contacts between nodes.

for a single node in the network. The investigation of network coding based DTN routing schemes has started with the simpler case where a single generation/block of (broadcast or unicast) packets is propagated in the network. With insights obtained from this simpler case, more realistic scenarios such as multiple unicast flows with a continuous packet arrival process have been studied [55, 57].

2.2. DTN Routing Schemes Overview

Recall that DTN routing adopts a so-called "store-carry-forward" paradigm. Under this paradigm, each node in the network *stores* a packet that has been forwarded to it by another node, *carries* the packet while it moves around, and *forwards* it to other relay nodes or the destination node when they come within transmission range. In this section, we provide an overview of the basic operations of broadcast and unicast routing schemes that have been proposed in the literature, and leave the detailed discussion of various design options to Section 3.

2.2.1 DTN Broadcast Routing Schemes

Broadcast has long been studied in the context of wireless *ad hoc* networks. The early broadcast routing scheme is based on flooding, where each node in the network rebroadcasts each message it receives. In relatively dense networks, this leads to excess overhead and contention [40]. It has been shown that the *gossip* based routing scheme [13], where each node rebroadcasts a message it received with a certain probability (i.e., *forwarding probability*), significantly reduces the overhead of the routing protocol, i.e., the number of messages sent in the network.

Under a DTN, at any point of time, a node might have no neighboring nodes, therefore, the broadcast routing scheme needs to decide *when* to rebroadcast a packet depending on the control signaling adopted (Section 3). Furthermore, as each rebroadcasting only reaches one or zero neighbors, each node needs to retransmit a packet multiple times so that a certain fraction of network nodes receives the packet. For these reasons, the *probabilistic routing* proposed in [49] extended the above gossiping scheme by generalizing the forwarding probability to the *forwarding factor*, f, that can take values larger than 1. If $f \leq 1$, a node rebroadcasts a copy of each packet it receives once with probability f; if $f > 1$, a node rebroadcasts the packet $\lfloor f \rfloor$ times, and rebroadcasts one more copy of the packet with probability $f - \lfloor f \rfloor$. Special care is taken at the source node so that a packet is transmitted at least once by the source, i.e., source node broadcasts a packet for $\max(1, \lfloor f \rfloor)$ times, and an additional copy is transmitted with probability $f - \lfloor f \rfloor$ if $f > 1$. The forwarding factor directly determines the forwarding overhead of the scheme, and clearly should be adjusted based on the node density.[1]

The performance metrics for evaluating different broadcast routing schemes ([11, 49]) include: (i) *energy efficiency*, often measured by the total number of transmissions required to transmit a packet to all receivers; (ii) *packet delivery ratio*, the fraction of packets that are received by all the destinations; and (iii) *packet delivery delay*.

2.2.2 DTN Unicast Routing Schemes

For unicast communication, each packet (generated by its source node) is destined to a single node (its destination node) in the network.

Unicast routing schemes for DTNs can be classified as single-copy or multicopy schemes. Under single-copy schemes [46], each packet is *forwarded* (*not copied*) along a single path, and at any point of time, there is a single copy of the packet in the network. The single-copy schemes incur a low transmission overhead and place minimal demand on the node

---

[1] Reference [11] proposed an adaptive scheme for each node to dynamically adjust the forwarding factor based on knowledge about its neighbors that are within two hops away.

buffer space. But when the future meeting process is not known in advance, forwarding decisions can later turn out to be wrong and in general lead to suboptimal performance. In such cases, it is often beneficial to use multi-copy schemes to reduce delivery delay and increase the delivery probability at the expense of larger transmission overhead and buffer occupancy. Under multi-copy schemes, a packet is *copied* (i.e., *duplicated*) to other nodes to be simultaneously forwarded along multiple paths to the destination; there are usually multiple simultaneous copies of a packet in the network at a given point of time. For example, the epidemic routing proposed by Vahdat and Becker [48] essentially floods the whole network in order to deliver a packet. By making use of all transmission opportunities, epidemic routing achieves minimum delivery delay when a network is lightly loaded, but causes resource contention when the network is loaded. Many variations of epidemic routing that trade-off delivery delay for resource consumptions have been subsequently proposed and studied, including K-hop, probabilistic forwarding [14] and spray-and-wait [44, 45, 47] schemes.

Under multi-copy routing schemes, when a packet is first delivered to the destination, there might be multiple copies of the packet in the network. *Recovery schemes* have been proposed to delete these obsolete copies from the network to free up storage space and avoid useless transmissions [14]. More details about various recovery schemes are presented in Section 3.

The performance metrics of interest for unicast applications are *packet delivery delay* and *the total number of copies or combinations made* for the packet in the network. The latter is a direct indication of the transmission overhead of a routing scheme. Clearly there exists an inherent trade-off between the packet delivery delay and the number of copies made [56].

2.3. Random Linear Coding

We now briefly describe the basic operation of Random Linear Coding (RLC), which is the technique considered by all existing works applying network coding to DTNs. For a more formal introduction to RLC, please refer to Chapter 1 of this book.

We assume that all packets are of the same length with P bits payload. When RLC is used in packet data networks, the payload of each packet can be viewed as a vector over a finite field [29], $\mathbb{F}_q$ of size q, more specifically, a packet of P bits is viewed as a $d = \lceil P/log_2(q) \rceil$ dimensional vector over $\mathbb{F}_q$.

A collection of packets that may be linearly coded together by network nodes is called a *generation*. Consider a generation with K original packets, $\mathbf{m}_i \in F_q^d, i = 1, 2, \ldots, K$. A linear combination of the K packets is:

$$\mathbf{x} = \sum_{i=1}^{K} \alpha_i \mathbf{m}_i, \ \alpha_i \in \mathbb{F}_q,$$

where the addition and multiplication operations are over $\mathbb{F}_q$. The coefficient $\alpha = (\alpha_1, \ldots, \alpha_K)$ is called the *encoding vector*, and the resulting linear combination, $\mathbf{x}$, is an *encoded message*. We say that two or more encoded messages are linearly independent if their encoding vectors are linearly independent. Each original packet, $\mathbf{m_i}$, can be viewed as a special combination with coefficients $\alpha_i = 1$, and $\alpha_j = 0, \forall j \neq i$.

Under an RLC-based routing scheme, networks nodes store and forward encoded messages together with their coefficients. For a generation of size K, the coefficients take up K symbols; while each data packet and combination takes up d symbols, with $d = \lceil P/log_2 q \rceil$, resulting in a relative overhead (i.e., the ratio of the size of the encoding coefficients and the data packet) of $K/(\lceil P/log_2(q) \rceil) \approx Klog_2(q)/P$. If in the set of encoded messages carried by a node there are at most r linearly independent encoded messages $\mathbf{x}_1, \ldots, \mathbf{x}_r$, we say that the rank of the node is r, and refer to the $r \times K$ matrix whose rows are the r encoding vectors as the node's *encoding matrix*, A. Essentially, this node has stored r independent linear equations with the K source packets as the unknown variables, i.e., $AM = X$, where $M = (\mathbf{m}_1, \ldots \mathbf{m}_K)^T$ is a $K \times 1$ matrix of the K original packets, and $X = (\mathbf{x}_1, \ldots \mathbf{x}_r)^T$ is the $r \times 1$ matrix of the r encoded messages. When a node (e.g., the destination) reaches rank K (i.e., full rank), it can decode the original K packets through matrix inversion, i.e., solve $AM = X$ for $M = A^{-1}X$ using a standard Gaussian elimination algorithm.[2]

We illustrate the data forwarding using the transmission from node u to node v as example. Node u generates a random linear combination (x_{new}) of its currently stored combinations, say $x_1, \ldots, x_r$: $x_{new} = \sum_{j=1}^{r} \beta_j x_j$, where the coefficients $\beta_1, \ldots \beta_r$ are chosen uniformly randomly from $\mathbb{F}_q$. Clearly,

---

[2] It is possible that the destination node decodes an original packet before the matrix reaches full rank, as long as the encoding matrix A contains a vector that has exactly one non-zero coefficient.

x_{new} is also a linear combination of the original K packets. This new combination, along with the coefficients *with respect to the original packets*, is forwarded to node v. If there is at least one combination stored in node u that cannot be linearly expressed by the combinations stored in node v, then node u has useful (i.e., *innovative*) information for node v, and the new random combination x_{new} is useful to node v (i.e., can increase the rank of node v) with probability greater or equal to $1 - 1/q$ [9].[3]

The RLC scheme incurs computation overhead as nodes perform random linear combinations and the destination node performs decoding operations. While the complexity of the encoding operation grows linearly with the generation size, the decoding operation has quadratic complexity in generation size.

3. DESIGN SPACE

Before presenting the main findings about the benefits of RLC in DTNs, we discuss the different design options for DTN routing schemes to explore the design space of DTN routing schemes. All these design options except generation management are applicable to both non-coding schemes and RLC schemes. All design options except recovery scheme are applicable to both broadcast and unicast schemes.

Generation Management

A coding based scheme needs to address the question of how many and which packets form a generation, i.e., generation management. Packets cannot be arbitrarily coded together for the following two reasons. First, the overhead of transmitting and storing encoding coefficients grows with the generation size, and so does the computational complexity of decoding algorithm. Secondly, for unicast applications, when K packets belonging to K different (unicast) flows are coded together, a destination has to receive K coded packets just to decode the one packet destined for it. The second consideration does not apply to broadcast communication, where all nodes are interested in receiving all packets.

---

[3] Knowing the encoding matrix of node v, node u can iteratively generate random linear combinations from its stored combinations until a combination useful to node v is generated. Alternatively, node u can also generate a combination useful to node v by using the deterministic algorithm proposed in [21]. Such processing trades off computational overhead for savings in transmission bandwidth.

Control Signaling

Because of the *ad hoc* nature and dynamically changing topology of DTNs, nodes perform beaconing in order to discover their neighbors (via broadcasting periodic beacon packets), and/or exchange with neighbors information about packets/coded packets they carried. Such control signaling is useful for nodes to decide *whether to transmit* and *what information to transmit*. The following different levels of control signaling have been considered in the literature:

- *No Signaling*: Under this most basic case (referred to as *no beacon* in [49]), no information about the neighborhood is available. Nodes decide to transmit packets without knowing whether there is a neighboring node or not.

- *Normal Signaling*: Under normal signaling (referred to as *normal beacon* in [49]), each node periodically transmits beacon messages in order to discover neighboring nodes, i.e., nodes within its transmission range. With normal signaling, a node typically only transmits information when it detects at least one neighbor.

- *Full Signaling*: Under full signaling (referred to as *intelligent beacon* in [49]), each node not only performs periodic beaconing to discover its neighbors, but also exchanges with its neighbors information about what packets or coded packets are stored locally, i.e., the sequence numbers of packets or the encoding vectors of coded packets. Based on such information, a network node typically only transmits to its neighbors if it has useful information for them.

Replication Control

For resource constrained DTNs where nodes have limited energy, or finite transmission bandwidth, or both, it is beneficial to control the total number of times that a packet (or a generation) is transmitted in the network, through a so-called *replication control* mechanism.

In the probabilistic routing scheme proposed in [49] for broadcast applications, the replication control is through the forwarding factor.

Spray-and-wait schemes ([45, 47, 55]) adopt a different replication control mechanism, where the total number of times a packet is transmitted in the network is directly controlled. Under the binary spray-and-wait scheme [45, 47], the source node assigns a counter value (a number of *tokens*), denoted as L, to each source packet it generates, which specifies the maximum number of copies that can be made for the packet in the network. When a node carrying a packet with token value l ($l > 1$) meets another relay node that does not carry a copy of the packet, the packet is forwarded to the latter node and the l tokens are equally split between the two copies of the packet.[4] A node carrying a packet with a token value of 1 does not forward the packet to relay nodes; it only delivers the packet to the destination. In this way, the total number of copies made for the packet in the whole network is bounded by L, though the actual number of copies being made is often smaller than L when a recovery scheme is employed. In Section 5.2, we review and compare two different replication control schemes based on the binary spray-and-wait scheme, that have been proposed to be used in conjunction with RLC.

Transmission Scheduling and Buffer Management

Routing schemes running on DTNs with resource constraints need to deal with resource contentions through transmission scheduling and buffer management [3, 28]. When a node encounters another node, the scheduling mechanism decides, among all candidate packets or generations in its buffer, which packets or generations to transmit to the other node. When a node with a full buffer receives a new original packet or coded packet, it decides whether and how to make space for the new packet based on the buffer management policy [15]. Different transmission scheduling and buffer management schemes result in different system performance, such as system wide average delivery delay.

In [31], the following different scheduling policies are considered for non-coding schemes:

- *random policy* that chooses each packet with the same probability;
- *local rarest policy* where each node chooses to transmit the packet that it has transmitted least up to that time instant;
- *global rarest policy* an oracle scheme where a node chooses the packet that has the smallest number of copies in the network.

Focusing on the potential benefits of network coding, [55, 57] have studied round robin scheduling for the source packets and randomized transmission scheduling for relay packets. This means that the source node

---

[4] If l is odd, the former copy keeps $\lceil l/2 \rceil$ tokens and the new copy is assigned $\lfloor l/2 \rfloor$ tokens.

takes turns to transmit each of the source packets during encounters,[5] and a relay node selects uniformly at random a packet or a generation to transmit during an encounter.

For the buffer management, a drophead scheme has been considered by [55, 57]: when a node with a full buffer receives a packet, it drops the relay packet[6] that has resided in the buffer the longest. Under an RLC scheme, when a node with a full buffer receives a combination, it randomly combines it with a combination existing in the buffer and replaces the existing combination with the new combination.

Recovery Scheme

For unicast applications, a multi-copy DTN routing scheme such as an epidemic routing and spray-and-wait scheme often employs a *recovery scheme* to save resource consumption [14, 56]. For example, under the *VACCINE* recovery scheme [14, 56], an anti-packet is generated by the destination when it first receives a packet, which is then propagated in the entire network, in the same fashion that a data packet is propagated under epidemic routing, to delete obsolete copies of the packet. Among the different recovery schemes, VACCINE recovery leads to the most significant resource savings, and therefore is adopted by existing works on network coding's benefit in DTN unicast application. In particular, [55] extended VACCINE (and any other) recovery scheme to work on RLC schemes: when a generation of packets is first delivered to its destination(s), acknowledgement information (called *anti-generation*) is generated by the destination(s) and propagated in the network to delete remaining copies of the packets or combinations of packets that belong to the generation.

4. CODING BENEFITS FOR BROADCAST COMMUNICATION

In [11, 49], the authors investigated the benefit of network coding for broadcast applications in wireless networks, considering different scenarios such as static or mobile wireless networks with different node densities. In

---

[5] With round robin scheduling at the source, all source packets are given equal opportunity to be disseminated into the network, yielding a smaller block delivery delay than the purely randomized scheduling [57].

[6] It is often assumed that network nodes have sufficient storage to store their own source packets.

this section, we present their results for sparse mobile wireless networks, a type of DTN. We first review the theoretical result in [11] which shows that RLC schemes require on average fewer transmissions to reach all the nodes, and then present simulation studies that examine the effects of replication control, control signaling level, and mobility models.

4.1. Coding Benefits in Energy Efficiency

For broadcast applications in DTNs, network coding based schemes deliver all messages to all nodes with fewer transmissions than non-coding schemes, and therefore improve energy efficiency [11].

Consider a network of N nodes, where each node has generated a packet to be broadcast to all the other nodes. Assume nodes move according to the *uniform at random* mobility model, i.e., at each time slot each node independently jumps to a new location in the terrain selected uniformly at random. At each time slot, each node decides to turn off or on its radio respectively, with probability p and $1 - p$. Assume there is no control signaling (no information about neighboring nodes and the information they carry). In each time slot, each node that is turned on randomly chooses a packet to transmit (under a non-coding scheme), or transmits a random linear combination of its coded packets to its neighbors (under an RLC scheme). There are on average $(1 - p)N$ transmissions in the network at each time slot.

Theorem 1. *[11] Broadcasting to all receivers can be achieved using on average the:*

- $T_w = \frac{N \log N}{(1-p)^2}$ *time slots, without using network coding;*
- $T_{nc} = \frac{\Theta(N)}{(1-p)^2}$ *time slots, using network coding with a large enough field size, q.*

Thus, on average, the ratio of the time slots needed to broadcast to all receivers without and with network coding is:

$$\frac{T_w}{T_{nc}} = \Theta(\log N).$$

Note that the ratio T_w/T_{nc} is equal to the ratio of the total number of transmissions required to broadcast to all the receivers in the two cases (without and with RLC), because the average number of transmissions at each time slot is same under the two schemes, i.e., $N(1 - p)$.

As observed in [11], the benefit of RLC in the above setting is similar in spirit to that of algebraic gossip, an RLC based protocol for message dissemination [9]. Both problems can be viewed as special instances of the *coupon collector's problem* that considers drawings with replacement from a set of N different coupons and studies the number of trials needed to obtain all N coupons. On the average, collecting all the coupons requires $O(N \log N)$ drawings [39]. Basically, as one has drawn more and more coupons, the probability that a new coupon is drawn in the following trial becomes smaller and smaller. By contrast, RLC schemes with a large enough field correspond to a modified version of the coupon collector's problem, where each new trial brings a new coupon with high probability, therefore one only needs $O(N)$ trials in order to collect all N coupons [9].

4.2. Practical RLC Broadcast Scheme

The performance comparison of simple routing schemes in the previous section considers the uniform at random mobility model, and assumes no control signaling or replication control mechanism. For more practical settings where nodes move according to a random waypoint mobility model, and furthermore, when a control signaling and replication control mechanism are employed, [11, 49] carried out simulation studies to evaluate the benefit of RLC schemes.

Under the uniform at random mobility model, nodes are completely reshuffled at each time slot. Common mobility models such as Random Waypoint Model (RWP) and random direction model exhibit similar memoryless property, when the transmission range is small in comparison to the region where nodes move and/or in comparison to the distance traveled by a node during a beaconing period (see Section 2.1). We expect the RLC scheme to provide similar benefits as under the uniform at random mobility model. However, when node movement is slower and the transmission range is larger, data packets are less well "mixed", leading to less significant benefits of RLC scheme. Figure 10.2 plots the ratios of the number of transmissions required by the non-coding and the RLC scheme under three different mobility settings. We observe that the relative gain of the RLC scheme under RWP is smaller than that under the uniformly random mobility model. In particular, the slower the node velocity, the smaller the gain, because the node shuffling is reduced.

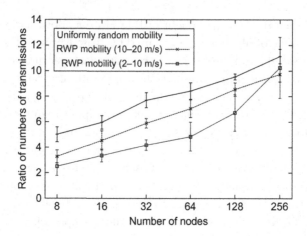

Figure 10.2 Ratios of transmissions required by simple flooding and an RLC broadcast scheme under a uniform at random mobility model and Random Waypoint Mobility (RWP, with different average velocity) (Fig. 10.8 in [11]).

We now consider the benefit offered by RLC when control signaling and replication control are introduced. Figure 10.3 plots the *packet delivery ratio* achieved by: (i) the probabilistic routing (i.e., non-coding) scheme; and (ii) the RLC scheme, when the amount of replication is controlled by the *forwarding factor* as introduced in Section 2.2.1. We observe that both non-coding and RLC schemes perform poorly under a static setting (i.e., the "Static Network" curve). On the contrary, when nodes have a RWP mobility with zero pause time and a minimum and maximum speed of 2 m/s and 10 m/s, respectively, the RLC scheme outperforms the non-coding scheme for all three control signalings. The benefits of network coding are remarkable when no signaling is performed or only normal signaling is adopted. For example, if no signaling is employed, while the non-coding scheme fails to deliver almost all packets with forwarding factor as large as 4, the RLC scheme delivers 80% of the packets with a forwarding factor of 4. When full signaling is used, both schemes achieve a 100% delivery ratio with a forwarding factor of 1. This demonstrates that RLC facilitates the design of an efficient, low-complexity, distributed broadcast routing protocol. We also observe that the RLC scheme provides a smoother trade-off between forwarding overhead and packet delivery ratio.

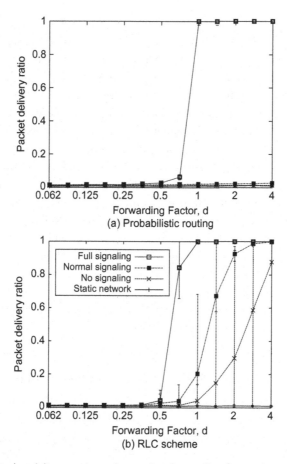

Figure 10.3 Packet delivery ratio under varying forwarding factors in sparse networks (Fig. 10.4 in [49]).

5. CODING BENEFITS FOR UNICAST APPLICATIONS

In this section, we present research contributions on the benefits of network coding for unicast applications in DTNs. The main focus is on the simple case where there is a single unicast flow made up of a block of K packets, propagating in a DTN where bandwidth and buffer are constrained. We use the 4-tuple (s, d, t_0, K) to denote a block of K unicast packets that is generated by source node s at time t_0, all of which are destined to the same destination node d. During each contact, b ($b < K$) packets can be exchanged. Note that without this limitation, the block could spread as a single message and there would be no need to perform network coding. Each node can carry B ($B < K$) *relay* packets but has enough buffer space to store packets originating from or destined to itself. The performance metrics of interest are the *block delivery delay*, i.e., the time to deliver the block of packets, and the *total number of copies or combinations made*.

In Section 5.1 we show that applying RLC to the block of packets reduces block delivery delay. Then in Section 5.2, we demonstrate that when a replication control scheme is employed, RLC schemes improve the delivery delay versus transmission number trade-off, and achieve smaller network wide average block delivery delay in multiple generation cases. We discuss how bandwidth and buffer constraints, block size, generation forming, and different control signaling levels affect the relative benefits and the overhead of RLC schemes in Section 5.3. In Section 5.4, we present a modeling study that characterizes the performance of an RLC scheme. Finally, Section 5.5 reviews other research works that applied network coding to DTN unicast communications.

5.1. Network Coding Reduces Block Delivery Delay

In this section, we demonstrate the benefits of RLC schemes in reducing block delivery delay. For a given block of packets, (s, d, t_0, K), and a fixed contact trace, there is a *minimum block delivery delay*, achievable by a centralized oracle scheme with knowledge of all future contacts, and a lower bound for block delivery delay achieved by any routing scheme (Section 5.1.1). Thanks to the increased randomness in data forwarding, RLC schemes achieve the minimum block delivery delay with high probability (Section 5.1.2). Finally, we discuss the performance of RLC schemes in terms of other metrics in Section 5.1.3.

5.1.1 Minimum Block Delivery Delay

For a block of packets specified by a tuple (s, d, t_0, K), i.e., a block of K packets generated by a source s at time t_0 and destined to a destination d under a given contact trace, the minimum block delivery delay can be calculated using an algorithm proposed in [57].

The algorithm constructs a time–independent *event-driven graph* based on the contact trace as follows [16] (Fig. 10.4 illustrates the event-driven graph for the contact trace in Fig. 10.1). Starting with an empty event-driven graph, the contact events in the contact trace are processed according to their time order. For a contact between mobile nodes i and j at time t, two nodes labeled as (i, t) and (j, t), and a link connecting them is inserted into the graph. The label of the link indicates the number of packets that can be transfered in each direction during the contact. Moreover, each of the

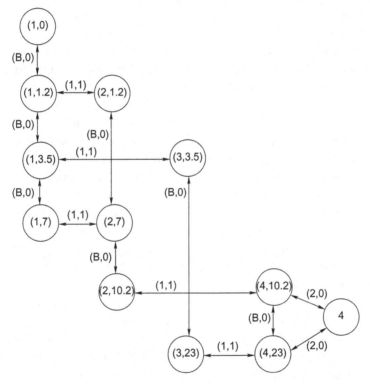

Figure 10.4 Event-driven graph for calculating minimum block delivery delay for block $(1, 4, 0, 2)$ under the contact trace shown in Fig. 10.1, assuming each node can store at most B relay packets. Links over which random linear combinations are transmitted are drawn in thick lines (there are $\eta = 4$ such links), e.g., over link $(1, 1.2) \rightarrow (2, 1.2)$, node 1 transmits a random linear combination of the two packets to node 2. Over link $(3, 23) \rightarrow (4, 23)$, node 3 does not perform RLC as it has only one encoded packet in its buffer.

two nodes $((i, t)$ and $(j, t))$ is connected to the node corresponding to the previous contact of the same mobile node (if any). The labels of these edges represent again the maximum number of packets that can be "transfered", the first value corresponds to the buffer size of the node, while the second value is always equal to 0.[7] For example, for the contact between node 1 and 3 at time 3.5, nodes $(1, 3.5)$ and $(3, 3.5)$ are inserted into the graph, with a link labeled with $(1, 1)$ connecting them. $(1, 3.5)$ is also connected to $(1, 1.2)$ with a link labeled as $(B, 0)$, given that each node can store B relay packets. In order to calculate the block delivery delay for a block of packets specified by tuple (s, d, t_0, K), two special nodes (s, t_0) and d are added into the graph (e.g., node $(1, 0)$ and 4 in Fig. 10.4).

The event-driven graph is a static graph that not only captures the temporal order of the contacts in the contact trace, but also represents the bandwidth and buffer constraints. The propagation of packets under the contact trace corresponds to the propagation of packets in this static graph: a mobile node (say node 1) transmitting a packet (coded or original) to another mobile node (say node 2) at a particular time instance $(t = 1.2)$ corresponds to the transmission of the packet over a horizontal link $((1, 1.2) \rightarrow (2, 1.2))$; a mobile node (say node 2) buffering a packet until a future contact (at $t = 7$) corresponds to the transmission of the packet over a vertical link (such as $(2, 1.2) \rightarrow (2, 7)$) in the static graph. As a result, classical graph algorithms can be applied to the graph [16] to solve various networking problems. In particular, [57] show how the event-driven graph can be advantageously used to calculate the minimum block delivery delay for (s, d, t_0, K). First, we observe that the existence of a flow of value k from (s, t_0) to d in the event-driven graph implies the existence of a set of forwarding decisions that enable the transfer of k packets from the source to the destination using the contacts that have been considered to generate the event-driven graph. In particular, if these contacts occur in the time range $[t_0, t]$, the existence of a flow of value k implies the possibility to transfer k packets generated at time t_0 by time t. Given these considerations, the minimum block delivery time for (s, d, t_0, K) is the earliest time t' for there to be a flow of value K from (s, t_0) to d in the event-driven graph built with the contacts in $[t_0, t']$. The minimum block delivery delay is the minimum block delivery time t' minus the initial time instant t_0.

---

[7] A packet received by node i at time t cannot be transfered to this node in the past.

The algorithm in [57] finds the minimum block delivery delay iteratively. Starting with an empty event-driven graph, the algorithm scans the contact trace from the block generation time, t_0, and gradually enlarges the graph by considering contact events according to their time order until a set of paths with a total capacity of K from the node (s, t_0) to the node d is found in the event-driven graph (using the Ford-Fulkerson algorithm [7]). Upon termination, the algorithm returns the time of the last contact considered, which is the minimum block delivery time.

5.1.2 Probability to Achieve Minimum Block Delivery Delay

We now consider how randomized non-coding and coding schemes perform in the practical setting where nodes have no knowledge about future contacts. We use a block of packets, $(1, 4, 0, 2)$, under the contact trace shown in Fig. 10.1 as an example. The minimum block delivery delay is 23 as it is only after the contact at time 23 is processed that the maximum flow from node $(1, 0)$ to node 4 in the event-driven graph (Fig. 10.4) reaches 2.

We first consider the non-coding scheme with random scheduling at relay nodes and round robin scheduling at source node, assuming full signaling is used. Node 1 first forwards m_1 to node 2 at time $t = 1.2$, then forwards m_2 to node 2 at time $t = 3.5$, and finally forwards m_2 to node 2 at time $t = 7$ (as node 2 already has m_1). When nodes 2 and 4 meet at $t = 10.2$, node 2 randomly selects a packet (m_1 or m_2) from its buffer to deliver to node 4. There are two possibilities:

(i) with probability 0.5, packet m_1 is selected to be delivered to node 4, and the minimum block delivery delay is achieved when node 3 delivers packet m_2 at $t = 23$;

(ii) with probability 0.5, packet m_2 is selected. As a result, when node 3 meets node 4 at $t = 23$, it has no useful information for node 4. Hence, the non-coding scheme achieves the minimum delay with probability 0.5.

Under the RLC scheme, node 1 forwards random linear combinations c_1 and c_2 to node 2, and c_3 to node 3 at times $t = 1.2, 7$, and 3.5, respectively. Node 3 transmits c_3 to node 4 at time $t = 23$. We distinguish two cases depending on whether c_1 and c_2 are independent or not.

(i) If c_1 and c_2 are independent, node 2 stores both combinations, and generates a random linear combination c_{12} of c_1 and c_2, and forwards it to node 4, when it meets node 4. As c_3 can be linearly expressed by c_1 and c_2 (given that c_1 and c_2 are independent), c_{12}, the random linear combination of c_1 and c_2, is independent of c_3 with probability $1 - 1/q$. We therefore conclude that for this case, with probability $1 - 1/q$, node 4 can decode the two original packets from c_{12} and c_3 at time $t = 23$.

(ii) If combinations c_1 and c_2 are linearly dependent, then node 2 only stores c_1, which is forwarded to node 4 at $t = 10.2$. Then, with probability $1 - 1/q$, c_3 and c_1 are independent, and node 4 reaches full rank at $t = 23$. Because the two cases are exclusive and exhaustive, we conclude that for this contact trace, the RLC scheme achieves minimum block delivery delay for block $(1, 4, 0, 2)$ with probability $1 - 1/q$. For $q = 2^8$, a commonly used finite field size, this probability is much larger than the probability of 0.5 achieved by the non-coding scheme.

From this example, we observe that RLC schemes provide much larger randomness than non-coding schemes, as the network nodes randomly and independently combine their stored coded packets to generate and forward coded packets, and the number of independent coded packets is much larger than the number of packets, K, from which non-coding schemes choose one to forward. As a result, the probability that a piece of redundant information is forwarded is much smaller for RLC schemes than for non-coding schemes.

In general, the probability that an RLC scheme achieves the minimum block delivery delay depends on the contact trace, more specifically, on the sequence of contacts that occur in the network after the block generation time. In order to have more insight into RLC schemes, we provide a different way to look at information propagation. Consider again the block $(1, 4, 0, 2)$ under the contact trace shown in Fig. 10.1. The RLC based DTN routing scheme corresponds to an RLC based transmission scheme on the corresponding static graph shown in Fig. 10.4. A transmission over a horizontal link such as link $(2, 10.2) \rightarrow (4, 10.2)$, corresponds to the transmission from a mobile node (node 2) to another mobile node (node 4) at the particular time instance ($t = 10.2$). Such transmission involves RLC operations if the sender node combines its stored packets before forwarding it to the receiver node. On the other hand, a transmission over a vertical link in the static graph represents data packets being buffered at the mobile node, and therefore does not involve RLC operation. The RLC scheme in the static graph achieves the maximum flow of $K = 2$ from node $(1, 0)$ to node 4 if, and only if, the RLC scheme delivers the block of $K = 2$ packets with the minimum block delivery delay in the DTN contact trace. The former problem is a special case of the general multiple-source multicast connection problem considered in [18]. This problem considers the transmission of a set of multicast packets originating from different source nodes to a common set of receiver nodes in a (static) graph $G = \langle V, E \rangle$ where each link has a certain capacity limit. If a given multicast flow, specified by the set of source nodes, the set of common receiver nodes, and the data rates of the sources, can be supported by the network, the following theorem in [18] provides a lower bound on the probability that the RLC scheme supports the multicast flow.

Theorem 2. *([18]) Consider a multicast connection problem on an arbitrary network with independent or linearly correlated sources, and a network code in which some or all network coding coefficients are chosen uniformly at random from a finite field $\mathbb{F}_q$ where $q > d$ (d is the number of multicast receivers), and the remaining code coefficients, if any, are fixed. If there exists a solution to the network connection problem with the same values for the fixed code coefficients, then the probability that the random network code is valid for the problem is at least $(1 - d/q)^\eta$, where η denotes the number of links associated with random coefficients.*

Note that η here corresponds to the number of links[8] along the set of source-to-destination paths (of total capacity K) that perform RLC operations. This theorem can be applied to our setting, by setting $d = 1$ for the unicast communication. Given that, by the minimum block delivery time, the static graph can support a flow of capacity equal to the size of the block, based on the above theorem, we conclude that the probability that the RLC scheme achieves the maximum flow value is at least $(1 - 1/q)^\eta$. In other words, the RLC scheme achieves the minimum block delivery delay with a probability at least $(1 - 1/q)^\eta$. This result demonstrates the underlying connection between the benefits of RLC in two seemingly different network settings, i.e., traditional static networks and DTNs with dynamic network topologies.

Simulation results have confirmed that the block delivery delay under the RLC scheme is very close to the minimum block delivery delay, as shown in Fig. 10.5(a), which plots the empirical cumulative distribution

---

[8] A link that can transmit $B > 1$ packets is counted as B links.

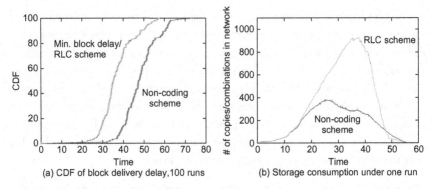

Figure 10.5 DTN with $N = 101$ nodes, homogeneous exponential inter-meeting time with rate $\beta = 0.0049$, bandwidth constraint of $b = 1$ packet per contact, and unlimited buffer space.

function (CDF) of minimum block delivery delay, and the block delivery delay achieved by the RLC and the non–coding scheme over 100 different simulation runs each with a different random seed.

5.1.3 Other Metrics

Having seen that the RLC scheme achieves smaller block delivery delay than the non–coding scheme, we now consider other performance metrics.

We first consider average packet delay and average in-order packet delay. Figure 10.6 plots the empirical CDFs of different delay metrics achieved by the RLC scheme and the non–coding scheme over 100 different simulation runs. We observe that, under the RLC scheme, all three metrics, i.e., block delivery delay, average delay, and in-order packet delay, are almost identical, and very close to the minimum block delivery delay. On the other hand, under the non–coding scheme, there is a significant difference between average packet delay, in-order packet delay, and block delivery delay. The RLC scheme performs better than the non–coding scheme in terms of block delivery delay and in-order packet delay (the improvement in terms of block delivery delay is larger than that of in-order packet delay), but performs worse in terms of average packet delay. Note that the RLC scheme considered here has some specific implementation peculiarities that improve its performance in terms of average delivery delay. For example, if a node can decode one or multiple packets before its matrix reaches full rank,

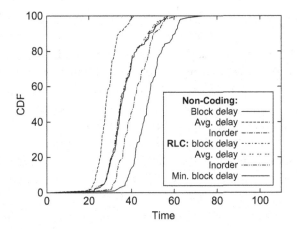

Figure 10.6 CDF of different delay metrics of 100 simulation runs, under the same network setting as that of Fig. 10.5.

it forwards the decoded packets (rather than random linear combinations of its coded packets) to the destination.

The faster information propagation of the RLC scheme is achieved at the price of more transmissions and larger buffer occupancy. For example, Fig. 10.5(b) plots the total number of packet copies (for the non-coding schemes) or combinations (for the RLC scheme) in the entire network as a function of time for one simulation run. Under the RLC scheme, network nodes randomly combine packets before forwarding them; as a result, when two nodes come into contact with each other, they are more likely to have useful information to exchange. This results in a sharper increase in the number of copies/combinations in the network under the RLC scheme. Furthermore, for the RLC scheme, the recovery process (VACCINE recovery is used here) starts only when the whole generation is delivered, whereas for the non-coding scheme, the recovery process for an individual packet starts immediately when the packet is delivered.

5.2. Network Coding Improves Delay vs. Transmission Number Trade-off

In this section, we show that when a replication control mechanism is employed to control the number of transmissions, the RLC scheme improves the delay versus number of transmission trade-offs, i.e., the RLC scheme achieves smaller average block delivery delay than the non-coding scheme with a similar number of transmissions in the network [55, 57].

We first consider the replication control mechanism proposed in [55, 57], the so-called *token-based* RLC scheme, which is based on the binary spray-and-wait scheme. In this scheme, the source node assigns a number of tokens (denoted as L_g) to each generation of packets that it generates, which limits the total number of combinations that can be exchanged for this generation in the network. Data forwarding follows the basic operation for RLC schemes described in Section 2.3, complemented with a specific token management policy. For the sake of simplicity, we consider a single generation, so that we can talk about the number of tokens and the rank of a node without the need to specify the generation. Consider, for example, a meeting between node u and node v. Node u is allowed to transmit a combination of packets that belongs to the generation if u carries a token number greater than 2. After the transmission, node u decrements its token number by 1 and the two nodes redistribute their token numbers, so that the sum of the two nodes' token numbers is reallocated to the two nodes in proportion to their rank, i.e., the number of linearly independent packets carried by the nodes.[9] A node carrying a token value of less than 2 can only transmit combinations of packets that belong to the generation for the destination. If the initial number of tokens for the generation is L_g, one can show that the total number of combinations transmitted for the generation is bounded by L_g. The actual number of combinations being transmitted is usually smaller than L_g when a recovery scheme is employed.

A different replication control approach for RLC, called E-NCP (Efficient Protocol based on Network Coding), was proposed in [31]. Under the E-NCP protocol, to transmit K source packets from a source node to a destination node, the source node generates K' (slightly larger than K) random linear combinations (which are referred to as pseudo source packets) from the K source packets, and disseminates these K' pseudo source

---

[9] Two nodes with no information to exchange also reallocate their token numbers in proportion to their rank, the rationale being to keep the potential of a node to spread information of the generation proportional to its rank.

packets to the first K' relay nodes it meets, respectively.[10] Each of the K' relay nodes subsequently uses a binary spray-and-wait mechanism to limit the total number of transmissions made for its pseudo source packet. Different pseudo source packets can then be combined at intermediate nodes (see details in [31]).

Simulation studies reported in [57] compared the block delivery delay versus transmission number trade-off achieved by the non-coding scheme with binary spray-and-wait applied to each of the K packets, the token-based RLC scheme and the E-NCP scheme. Figure 10.7 plots the average block delivery delay versus number of transmissions, for a block of K=10 packets, under different token limits, for the cases both without buffer constraints (a) and with buffer constraint of $B = 2$ (b). We observe that, with a similar number of transmissions, the RLC schemes achieve smaller block delivery delay than non-coding schemes, and the token-based RLC scheme outperforms the E-NCP scheme, especially for small numbers of transmissions. The results for a limited relay buffer case further establish the benefits of the RLC schemes in reducing block delivery delay without increasing transmission overheads.

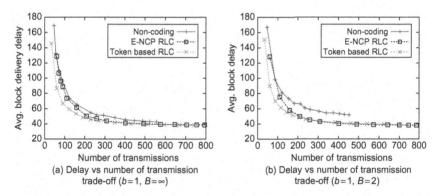

Figure 10.7 Block delivery delay vs transmission number trade-off under the same network setting as Fig. 10.5 except for the bandwidth and buffer constraints.

---

[10] The reason for disseminating K' pseudo source packets is so that the original K source packets can be decoded with high probability when K coded packets are received. As [57] pointed out, it suffices for the source to disseminate to the K relay nodes the original K source packets or K linearly independent coded packets.

The fact that the RLC scheme improves the delivery delay versus the number of transmission trade-offs explains the benefits observed for the RLC scheme under the multiple generation case [55, 57]. Consider a network scenario where there are multiple continuous unicast flows, each source independently generates blocks of packets according to a Poisson process, and RLC is applied to packets belonging to the same block, i.e., each block forms a generation. Whether network coding is used or not, a replication control mechanism, by limiting the number of transmissions made, can reduce bandwidth contention and in turn reduce the system-wide average block delivery delay. There exists an optimal token value under which the average block delivery delay is minimized: a too large token value or no replication control leads to severe contention in the network and degraded performance, while a too small token value prevents network nodes from exploiting all available bandwidth. In [55, 57] it has been shown that the minimum delivery delay achieved by the RLC scheme is smaller than that achieved by the non–coding scheme under various traffic rates, especially when the buffer space is constrained.

5.3. Discussion about RLC Benefits

In this section, we summarize the simulation studies reported in [55, 57] that illustrate the impact of different system parameters on the benefits of the RLC scheme.

5.3.1 Impact of Different Bandwidth and Buffer Constraints

For DTNs with bandwidth and buffer constraints, we have seen that the RLC scheme achieves much smaller block delivery delay than the non–coding scheme. As the network bandwidth becomes less and less constrained, the RLC benefit diminishes and becomes non-existent when the number of packets that can be exchanged during each contact, b, equals the block size K. In this case, the K packets propagate independently without competing for bandwidth.[11] For example, Fig. 10.8(a) from [57] plots the average block delivery delay and its 95% confidence interval (based on 50 different simulation runs) under varying bandwidth constraints, for the case

[11] Therefore, the block delivery delay coincides with the epidemic routing delay under no resource constraints as characterized in [56].

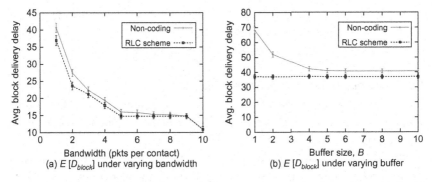

Figure 10.8 Impact of bandwidth and buffer constraints under the same network setting as Fig. 10.5 except the bandwidth and buffer constraint for (a) and (b) respectively.

where there are $k = 10$ packets from the same unicast flow propagating in a network of 101 nodes. Likewise, the improvement on the delay versus transmission trade-off achieved by the RLC scheme over the non-coding scheme diminishes when bandwidth becomes less constrained.

The RLC scheme is especially beneficial for the case where, in addition to the bandwidth constraint, the buffer space at relay nodes is also constrained. Figure 10.8(b) plots the average block delivery delay (and the 95% confidence interval) for a block of $K = 10$ packets achieved by the RLC scheme and the non-coding scheme under varying nodal buffer sizes, B ($B \leq K$). We observe that, as the buffer space becomes more and more constrained, the performance under the RLC scheme only degrades slightly, in sharp contrast to the non-coding scheme. As different packets are mixed randomly under the RLC scheme, dropping a combination has the same effects on all packets encoded in the combination. Therefore, the RLC scheme allows an even propagation of different information in the network. For the non-coding scheme, when a block of packets starts being spread in the network, small differences in the number of copies are amplified: the more copies a packet has in the network, the more this packet is copied to other nodes.[12] This results in an uneven propagation of different

---

[12] As a first approximation the spread rate of a packet with n copies in a network with N nodes is proportional to $n(N - n)$, then the packet with a larger number of copies spreads faster as long as $n < N/2$.

packets: some packets spread quickly to a large number of nodes, while others spread much more slowly. It therefore takes much longer to deliver the "slowest" packet, and therefore the whole block of packets.

5.3.2 Impact of Generation Management

We now discuss the impact of generation management, i.e., the decision on how many and which packets form a generation to which random linear coding operations are applied.

So far, we have focused on the case where a generation is formed by packets from the same unicast flow. Other ways to form a generation have been explored in [55, 57], including the case where multiple packets from different sources, but destined for the same destination, form a generation, and the case where multiple packets from different sources destined for different destinations form a generation. We refer to these three cases respectively as SS_SD, MS_SD, and MS_MD (Single/Multiple Source, Single/Multiple Destination). The benefit achieved by the RLC scheme for the MS_SD case is smaller than for the SS_SD case. Basically, under the MS_SD case, the K packets start to propagate from the K different source nodes, and under non-coding schemes the effect of relay nodes choosing the wrong packets to forward becomes less significant. For the MS_MD case, the RLC scheme performs worse than the non-coding scheme, as mixing packets destined for different destinations forces every destination node to receive K independent combinations to decode the one single packet destined to it. In general, the RLC scheme should restrict mixing to packets that are destined to the same destination.

References [55, 57] also show that as the generation size, K, increases, the relative benefit of the RLC schemes decreases. This is because, for the non-coding scheme, with a larger block size, there are a larger number of packets to randomly choose from, and therefore the probability of two paths choosing to forward the same packet is smaller.

5.3.3 Impact of Control Signaling

All above results for unicast application are for the *full signaling* case, where two encountering nodes first exchange information about which packets (or coded packets) they each carry in order to avoid transmitting useless information to each other. However, the exchange of this control information consumes transmission bandwidth, especially for RLC schemes, where more information is exchanged (the encoding coefficients for each coded packet compared to the packet ID for each packet). Also, for RLC schemes full signaling incurs computation overhead, because each node needs to perform some calculations in order to determine whether it has useful information for the other on the basis of the received encoding coefficients.

Now let us consider *normal signaling*, where nodes do not exchange information about the packets or coded packets they carry. Without such information, under the non-coding scheme, a node randomly chooses a packet from the set of packets it carries, and forwards it to the other node; under the RLC scheme, a node generates and transmits a random linear combination to the other node as long as the other node has not reached full rank yet. Figure 10.9 plots the block delivery delay versus the number of transmission trade-offs achieved by the non-coding and the RLC scheme under full signaling and normal signaling. We observe that the performance of the RLC scheme under normal signaling is almost identical to that under full signaling, whereas for the non-coding scheme, the performance under normal signaling is significantly worse than that under full signaling. This demonstrates that the RLC scheme retains its benefit over the non–coding

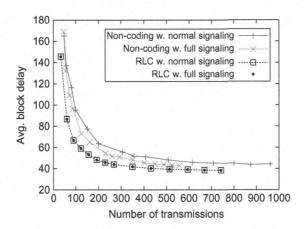

Figure 10.9 Block delivery delay vs transmission number trade-off with full signaling and normal signaling, the network setting is the same as that of Fig. 10.5.

scheme even with less signaling overhead, i.e., when normal signaling is used for the RLC scheme while full signaling is used for the non-coding scheme.

5.4. Modeling Studies of Network Coding Scheme

In order to quantitatively analyze the benefit of RLC schemes, an Ordinary Differential Equations (ODEs) based model has been proposed to characterize the block delivery delay under the non-coding and RLC scheme in [32, 33].

The model characterizes the propagation of a block of K packets that are generated by the source at time $t = 0$, all destined for the same destination node in a network with $N + 1$ mobile nodes. The inter-contact time between each pair of nodes follows an exponential distribution with rate λ. During each contact, one packet can be transmitted between two nodes in each direction. Each node has limited buffer space and can store at most B $(B < K)$ relay packets. For the related problem of gossip algorithm for spreading multiple rumors or messages to the whole network, many efforts have been focused on characterizing the stopping time, i.e., the number of time slots by which all nodes receive all messages achieved by different gossip algorithm ([37] and references therein).

Under simplifying assumptions to be explained later, for both the RLC scheme and the non-coding scheme, the system state at time t can be described by a B-tuple $\{X_1(t), X_2(t), \ldots, X_B(t)\}$, where $X_i(t)$ denotes the number of rank i nodes in the network (excluding the source node). The number of nodes with rank zero is given by $X_0(t) = N - \sum_{j=1}^{B} X_j(t)$.

Consider the RLC scheme with normal signaling[13] and further assume that, when a node receives a random linear combination from another node, its rank always increases by 1, i.e., the combination contains useful information for the node. This is an optimistic assumption, since the other node might have no useful information for the node,[14] and even if the other node has useful information for the node, the random linear

---

[13] This means that when two nodes, u and v, encounter each other, node u transmits a random linear combination to node v as long as u has rank greater than 0 and v has not reached full rank yet (the transmission from node v to node u is similar).

[14] i.e., all the coded packets at another node are linear combinations of the coded packets at the node.

combination it generated might contain no useful information for the node. The following ODEs model results:

$$\frac{dX_1(t)}{dt} = \lambda \left(\sum_{j=1}^{B} X_j(t) + 1 \right) X_0(t) - \lambda \sum_{j=1}^{B} X_j(t) X_1(t) \tag{1}$$

$$\frac{dX_i(t)}{dt} = \lambda \sum_{j=1}^{B} X_j(t) X_{i-1}(t) - \lambda \sum_{j=1}^{B} X_j(t) X_i(t)$$

$$\text{for } i = 1, \dots, B - 1, \tag{2}$$

$$\frac{dX_B(t)}{dt} = \lambda \sum_{j=1}^{B} X_j(t) X_{B-1}(t) \tag{3}$$

with initial conditions $X_i(t) = 0$ for $i = 1, \dots, B$. These ODEs characterize the changing rates of $X_i(t)$, $i = 1, 2, \dots, B$. For example, $X_1(t)$ increases by one whenever a rank 0 node encounters a node with rank greater than 0 (there are a total of $\sum_{j=1}^{B} X_j(t) + 1$ such nodes including the source node), therefore $X_1(t)$ increases with rate $\lambda \left(\sum_{j=1}^{B} X_j(t) + 1 \right) X_0(t)$. On the other hand, $X_1(t)$ decreases by one whenever a rank 1 node increases its rank by 1 by meeting some node (including the source node, but rather than itself) with rank greater than 0 (there are a total of $\sum_{j=1}^{B} X_j(t)$ such nodes), therefore the decrement rate of $X_1(t)$ is $\lambda \sum_{j=1}^{B} X_j(t) X_1(t)$. Considering both increasing and decreasing rates of $X_1(t)$ yields (1).

Let T_i, $i = 1, \dots, K$ denote the time instant when the destination node reaches rank i. The block delivery delay is then T_K. The following ODEs hold for the cumulative distribution function of T_i, $F_i(t) := \Pr(T_i < t)$:

$$\frac{dF_1(t)}{dt} = \lambda \left(\sum_{j=1}^{B} X_j(t) + 1 \right) (1 - F_1(t))$$

$$\frac{dF_i(t)}{dt} = \lambda \sum_{j=1}^{B} X_j(t)(F_{i-1}(t) - F_i(t)), \text{ for } i = 2, \dots, K,$$

with initial conditions given by $F_i(0) = 0$, for $i = 1, 2, \dots, K$.

For the non-coding scheme, in order to characterize the probability that two encountering nodes have useful information (i.e., new packets) for each other, it is assumed that the i packets (with $i = 1, \ldots, B$) carried by a rank i node have been drawn with equal probability from the original set of K packets. As a result, $\Pr(i,j)$, the probability that a node carrying i packets can receive a new packet from a node carrying j packets, can be expressed as:

$$\Pr(i,j) = \begin{cases} 1, & \text{if } i < j, \\ 1 - \binom{i}{j}/\binom{K}{j}, & \text{if } i \geq j. \end{cases}$$

The ODEs in Equations (1–3) can then be modified by introducing $\Pr(i,j)$ in order to model the non-coding scheme. Basically, the transition rates are "thinned" as the following example demonstrates:

$$\frac{dX_i}{dt} = \lambda \sum_{j=1}^{B} \Pr(i-1,j) X_j(t) X_{i-1}(t)$$

$$- \lambda \sum_{j=1}^{B} \Pr(i,j) X_j(t) X_i(t) \text{ for } i = 1, \ldots, B-1$$

The above ODEs are solved numerically to obtain the average block delivery delay of both the non-coding scheme and the coding scheme, and compared against simulation studies in [32, 33]. Three different scheduling policies for the non-coding scheme are simulated: *random policy*; *local rarest policy*; and *global rarest policy*. The average block delivery delay under the simulation is then compared against those predicted by the ODE models, as shown in Fig. 10.10, which plots the block delivery delay under varying N, the total number of nodes in the network excluding the source node. We observe that the ODE models underestimate the block delivery delay for both the RLC scheme and the non-coding scheme, an expected result considering the optimistic assumptions made. Among the three scheduling policies, global rarest policy, an oracle scheme assuming global knowledge about the number of copies each packet has in the network, performs the best, followed by the random policy, and the local rarest policy performs worst. The simulation results also show that the RLC scheme, which does

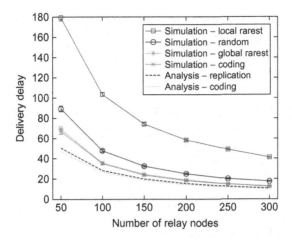

Figure 10.10 Block delivery delay under different number of nodes (Fig. 10.3 in [31]).

not rely on any oracle, performs as well as the non-coding scheme with global rarest policy, an oracle-based scheme.

5.5. Other Works on an RLC Scheme in Unicast Application

In this section, we briefly review other research efforts that apply network coding to unicast application in DTNs.

5.5.1 Priority Coding Protocol

We have seen that the RLC scheme performs better than the non-coding scheme in terms of time to deliver the last packet in the block, i.e., block delivery delay, but performs worse in terms of average packet delivery delay (Fig. 10.5(b)). For applications where different packets have different priorities, [31] proposes a priority coding protocol which works as follows. Suppose that the source generates K packets destined to the same destination node, and the K packets can be classified into M ($M < K$) priority levels, so that the packets of level i should be decoded before packets of level j, if $i < j$. The source node first transmits packets from level 1 using the RLC scheme, applying coding to all packets of this level. Only after receiving the ACK for the level 1 packets (generated by the destination and propagated by all nodes under VACCINE recovery scheme), does the source start to transmit packets from level 2 using the RLC scheme. Nodes in the network drop locally stored combinations of lower level packets when they receive combinations of higher level packets. The source starts to transmit packets of level $i + 1$ ($i = 1, 2 \ldots M - 1$) upon receiving ACK for packets of level i, until all K packets are decoded by the destination.

Using the ODE models in Section 5.4, [32] found that the overhead of this priority coding protocol, in terms of the increase in the time to deliver all the K packets as compared to the non-priority RLC scheme, increases linearly with the number of priority levels. For each priority level, the priority coding protocol incurs an extra delay equal to the time to deliver the first packet of the level, and the time to propagate ACK from the destination to the source.

5.5.2 Optimal Control of a Two-Hop Scheme

In [2], the authors consider the case where K packets are generated at a source (at different time instances) and need to be delivered to a destination by a deadline. The source can perform random linear combinations of the packets before forwarding information to relay nodes. Each relay node can store at most one linear combination, and only transfers the (coded) packet it carries to the destination. The purpose is to maximize the delivery probability of the K packets by the given deadline. The authors show that the best strategy for the source when meeting a relay is to combine all the packets available at that time. A lower bound for the delivery probability is also derived.

We comment that this routing scheme is essentially an RLC scheme with two-hop routing where each packet or coded packet can traverse a path of one- or two-hops to reach the destination. As only the source node performs random linear coding, it is not "network coding" in the strict sense.

5.5.3 Network Coding Based Secure Communication for DTN

For DTNs with unidentified Byzantine adversarial nodes that introduce corrupted packets into the network, an RLC based secure unicast routing protocol has been proposed in [42].

Consider a block of K packets generated at a source node and destined for a destination node. Suppose that the adversarial nodes can pollute the entire coded packets, i.e., including the encoding coefficients in the headers. The destination node might not be able to decode the K packets with any K independent coded packets, as some of them might be polluted.

The proposed protocol combines cryptographic key dissemination and error-correction coding together to ensure secure data transmission. The source node encodes the K data packets into the error-correction coded vector space [27], and then generates private and public keys from the coded-packets [58]. The public keys are then distributed for a certain fraction of nodes in the network. Nodes that have received the public keys are secured, as they can use the public keys to verify the received linear combinations. Finally, the source node sends the coded packets using the RLC scheme, adopting the E-NCP replication control mechanism.

For the simple setting of a single generation and no bandwidth or buffer constraints, the analysis of the above protocol [42] characterizes the probability that the destination node successfully decodes the K packets within a certain time, in terms of protocol parameters such as the dimension of the error-correction coded vector space, the number of nodes that need to receive the public keys, and the binary spray-and-wait counter.

6. OPEN ISSUES

In this section, we discuss several open issues that deserve further investigation.

6.1. RLC Benefits for Application with Short Messages

Until now, most existing works have applied random linear coding to a block of K packets that are generated at the source node in a batch, and have considered the time to deliver the whole block of packets (i.e., the block delivery delay). The batch arrival of multiple packets can be due to the applications generating long messages, e.g., in the case of file sharing. The long messages need to be fragmented into smaller packets, in part to take advantage of the often short transmission opportunities.

For applications that generate short messages, it is inefficient to fragment the short message into very small packets. More packets to send means more lower layer protocol overhead. Furthermore, the relative overhead of the RLC scheme (to store and transmit coding coefficients) as given by $Klog_2(q)/P$ is larger for shorter messages, i.e., for small values of P.

Whether and when random linear coding, or more generally network coding, is beneficial in this case remains an open question. Instead of block delivery delay, average packet delivery delay or in-order delivery delay should be considered, given that the packets are independent from each other. The central question to the operation of an RLC scheme is the generation forming, i.e., how to decide which packets form a generation to which RLC is applied. Above, we have seen that combining packets belonging to different unicast flows results in smaller benefits (nevertheless, it is still beneficial in the buffer constrained case). The options of combining packets (from the same unicast flow) generated at different times remain to be studied.

6.2. An RLC Scheme for Multicast Communication

Multicast communication supports the distribution of a data packet to a group of users. The application of network coding to multicast communication in DTNs has received less attention than that of broadcast and unicast, partly due to the different semantics and multicast algorithms for multicast in DTNs [1, 6, 54, 59].

For DTN based content-distribution networks with bandwidth constraints, reference [23] proposed adapting network coding to the different social interests of users, and only mix contents of a similar type. Preliminary simulation results show that a network coding based scheme achieves lower decoding delay than a non-coding scheme. Moreover, when network coding is only applied to messages of the same type, users that are only interested in a small set of contents experience smaller delays, at the cost of larger delays to users with wide interests. To trade-off between the potential throughput improvement and increased delay of the network coding based scheme, the authors proposed adjusting the mixing so that content of different types but with a large overlap in their receivers could be mixed.

Given that RLC schemes have shown better performance in the two extreme cases of multicast communication, i.e., broadcast and unicast, some results reviewed in this chapter might also apply to multicast communication. Further work needs to be done to fully understand the network coding's benefit for multicast communication in DTNs.

7. SUMMARY AND CONCLUSIONS

In this chapter, we reviewed existing research that applies network coding to Delay Tolerant Networks (DTNs). We first provided a background on traditional non-coding based routing schemes for broadcast and unicast applications in DTNs and on the basic operation of Random Linear Coding (RLC), the form of network coding usually considered in literature. Then we introduced the design space for DTN routing, emphasizing the different options to integrate network coding. The main focus of the chapter is on the performance evaluation of RLC based routing schemes both for broadcast and unicast communications, unicast communications having been the object of a larger amount of research. We highlighted both theoretic results and simulation studies findings. For both communication models, the RLC based scheme provides better trade-off between energy consumption and delivery performance.

Beyond the above results, we also reviewed various research efforts that extend the basic RLC schemes, for example to achieve priority routing and secure communication. Finally, we discussed open issues that require, in our opinion, further investigation.

REFERENCES

[1] M. Abdulla and R. Simon. A Simulation Analysis of Multicasting in Delay Tolerant Networks. In *Conference on Winter simulation (WSC)*, Monterey, CA, USA, 2006.
[2] E. Altman, F. De Pellegrini, and L. Sassatelli. Dynamic Control of Coding in Delay Tolerant Networks. In *IEEE International Conference on Computer Communications (INFOCOM)*, Shanghai, China, 2010.
[3] A. Balasubramanian, B. N. Levine, and A. Venkataramani. DTN Routing as a Resource Allocation Problem. In *ACM Conference on Communications Architectures, Protocols and Applications (SIGCOMM)*, 2007.
[4] J. Burgess, B. Gallagher, D. Jensen, and B. N. Levine. MaxProp: Routing for Vehicle-Based Disruption-Tolerant Networks. In *IEEE International Conference on Computer Communications (INFOCOM)*, 2006.
[5] T. Camp, J. Boleng, and V. Davies. A Survey of Mobility Models for *Ad Hoc* Network Research. In *Wireless Communications and Mobile Computing (WCMC): Special issue on Mobile Ad Hoc Network Research: Research, Trends and Applications*, 2002.
[6] M. C. Chuah and Y. Xi. An Encounter-based Multicast Scheme for Disruption Tolerant Networks. Technical Report LU-CSE-07-009, Lehigh University, 2007.
[7] T. H. Cormen, C. E. Leiserson, and R. L. Rivest. *Introduction to Algorithms*. The MIT Press, 1996.
[8] First Mile Solutions (i.e., DakNet). http://firstmilesolutions.com/.

[9]   S. Deb, M. Médard, and C. Choute. Algebraic Gossip: A Network Coding Approach to Optimal Multiple Rumor Mongering. *IEEE/ACM Transactions on Networking, special issue on networking and information theory*, pages 2486–2507, 2006.

[10] A. Doria, M. Ud'en, and D. P. Pandey. Providing Connectivity to the Saami Normadic Community. In *International Conference on Open Collaborative Design for Sustainable Innovation (dyd)*, Bangalore, India, December 2002.

[11] C. Fragouli, J. Widmer, and J.-Y. Le Boudec. Efficient broadcasting using network coding. *IEEE/ACM Transactions on Networking*, 16(2): 450–463, April 2008.

[12] R. Groenevelt, P. Nain, and G. Koole. The message delay in mobile *ad hoc* networks. *Performance Evaluation*, 62(1-4): 210–228, October 2005.

[13] Z. Haas, J. Halpern, and L. Li. Gossip-based *Ad Hoc* Routing. In *IEEE International Conference on Computer Communications (INFOCOM)*, New York, NY, USA, 2002.

[14] Z. J. Haas and T. Small. A new networking model for biological applications of *ad hoc* sensor networks. *IEEE/ACM Transactions on Networking*, 14(1): 27–40, February 2006.

[15] B. Haeupler. Analyzing Network Coding Gossip Made Easy (simpler proofs for stronger results even in adversarial dynamic networks). In *Allerton Conference on Communication, Control and Computing*, Monticello, IL, USA, 2010.

[16] D. Hay and P. Giaccone. Optimal routing and scheduling for deterministic delay tolerant networks. In *International Conference on Wireless On-Demand Network Systems and Services (WONS)*, Snowbird, Utah, USA, 2009.

[17] T. Ho, R. Koetter, M. Médard, D.R. Karger, and M. Effros. The Benefits of Coding Over Routing in a Randomized Setting. In *IEEE International Symposium on Information Theory (ISIT)*, Yokohama, Japan, June–July 2003.

[18] T. Ho, M. Médard, R. Koetter, and D.R. Karger. A random linear network coding approach to multicast. *IEEE Transactions on Information Theory*, 52(10): 4413–4430, 2006.

[19] P. Hui, A. Chaintreau, R. Gass, J. Scott, J. Crowcroft, and C. Diot. Pocket Switched Networking: Challenges, Feasibility, and Implementation Issues. In *IFIP TC6 International Workshop on Autonomic Communication (WAC)*, 2005.

[20] B. Hull, V. Bychkovsky, Y. Zhang, K. Chen, M. Goraczko, A. K. Miu, E. Shih, H. Balakrishnan, and S. Madden. CarTel: A Distributed Mobile Sensor Computing System. In *ACM Conference on Embedded Networked Sensor Systems (SenSys)*, New York, NY, USA, 2006.

[21] S. Jaggi, P. Sanders, P. A. Chou, M. Effros, S. Egner, K. Jain, and L. M. G. M. Tolhuizen. Polynomial time algorithms for multicast network code construction. *IEEE Transactions on Information Theory*, 51(6): 1973–1982, 2005.

[22] P. Juang, H. Oki, Y. Wang, M. Martonosi, L.-S. Peh, and D. Rubenstein. Energy-Efficient Computing for Wildlife Tracking: Design Trade-offs and Early Experiences with ZebraNet. In *ACM International Conference on Architectural Support for Programming Languages and Operating Systems (ASPLOS)*, San Jose, CA, USA, 2002.

[23] G. Karbaschi and A. C. Viana. A Content-based Network Coding to Match Social Interest Similarities in Delay Tolerant Networks. In *First Extreme Workshop on Communication (ExtremeCom)*, Laponia, Sweden, 2009.

[24] S. Katti, D. Katabi, W. Hu, and R. Hariharan. The Importance of Being Opportunistic: Practical Network Coding For Wireless Environments. In *Allerton Conference on Communication, Control, and Computing*, Montecello, IL, USA, 2005.

[25] S. Katti, H. Rahul, W. Hu, D. Katabi, M. Medard, and J. Crowcroft. XORs in the air: Practical wireless network coding. *IEEE/ACM Transactions on Networking*, 16(3): 497–510, 2008.

[26] D. Kempe, J. Kleinberg, and A. Kumar. Connectivity and inference problems for temporal networks. In *Journal of Computer and System Sciences, Special issue on STOC 2000*, 64(4): 820–842, 2002.

[27] R. Koetter and F. Kschischang. Coding for errors and erasures in random network coding. *IEEE Transactions on Information Theory*, 54(8): 3579–3591, August 2008.

[28] A. Krifa, C. Barakat, and T. Spyropoulos. Optimal buffer management policies for delay tolerant networks. In *IEEE Conference on Sensor, Mesh and Ad Hoc Communications and Networks (SECON)*, San Diego, CA, USA, 2007.

[29] R. Lidl and H. Niederreiter. *Finite Fields, 2nd edition*. Cambridge, England: Cambridge University Press, 1997.

[30] Y. Lin, B. Li, and B. Liang. Differentiated Data Persistence with Priority Random Linear Codes. In *International Conference on Distributed Computing Systems (ICDCS)*, Toronto, Canada, 2007.

[31] Y. Lin, B. Li, and B. Liang. Efficient network coded data transmissions in disruption tolerant networks. In *IEEE International Conference on Computer Communications (INFOCOM)*, Phoenix, AZ, USA, 2008.

[32] Y. Lin, B. Li, and B. Liang. Stochastic analysis of network coding in epidemic routing. *IEEE Journal on Selected Areas in Communications, Special Issue on Delay and Disruption Tolerant Wireless Communication Systems*, 26(5): 794–808, June 2008.

[33] Y. Lin, B. Liang, and B. Li. Performance modeling of network coding in epidemic routing. In *ACM/SIGMOBILE International Workshop on Mobile Opportunistic Networking (MobiOpp)*, Puerto Rico, 2007.

[34] J. Liu, D. Goeckel, and D. Towsley. Bounds on the gain of network coding and broadcasting in wireless networks. In *IEEE Conference on Computer Communications (INFOCOM)*, Anchorage, AK, USA, 2007.

[35] D. S. Lun, M. Médard, T. Ho, and R. Koetter. Network Coding with a Cost Criterion. In *International Symposium on Information Theory and its Applications (ISITA)*, Parma, Italy, 2004.

[36] D. S. Lun, N. Ratnakar, M. Médard, R. Koetter, D. R. Karger, T. Ho, E. Ahmed, and F. Zhao. Minimum-Cost Multicast over Coded Packet Networks. *IEEE Transactions on Information Theory*, 52(6): 2608–2623, 2006.

[37] C. Avin, M. Borokhovich, and Z. Lotker. Tight Bounds for Algebraic Gossip on Graphs. *IEEE International Symposium on Information Theory*, Austin, TX, 2010.

[38] A. Maffei, K. Fall, and D. Chayes. In *Ocean Instrument Internet: Using Disruption Tolerant Networking to Join Heterogeneous Oceanographic Instrumentation into a Single Network*, Honolulu, HI, USA, 2006.

[39] M. Mitzenmacher and E. Upfal. *Probability and computing: Randomized algorithms and probabilistic analysis*. Cambridge Press, 2005.

[40] S.-Y. Ni, Y.-C. Tseng, Y.-S. Chen, and J.-P. Sheu. The Broadcast Storm Problem in a Mobile *Ad Hoc* Network. In *ACM Internation Conference on Mobile Computing and Networking (Mobicom)*, Seattle, WA, USA, 1999.

[41] J. Partan, J. Kurose, and B. N. Levine. A Survey of Practical Issues in Underwater Networks. In *ACM International Workshop on UnderWater Networks (WUWNet)*, Los Angeles, CA, USA, 2006.

[42] L. Sassatelli and M. Médard. Network coding for delay tolerant networks with byzantine adversaries. Preprint 2009, arXiv:09075488, http://cdsweb.cern.ch/record/1195977/.

[43] T. Small and Z. J. Haas. The Shared Wireless Infostation Model – A New *Ad Hoc* Networking Paradigm. In *ACM International Symposium on Mobile Ad Hoc Networking and Computing (MOBIHOC)*, 2003.

[44] T. Small and Z. J. Haas. Resource and performance trade-offs in delay-tolerant wireless networks. In *SIGCOMM Workshop on Delay Tolerant Networking (WDTN)*, 2005.

[45] T. Spyropoulos, K. Psounis, and C. Raghavendra. Efficient routing in intermittently connected mobile networks: The multiple-copy case. In *ACM/IEEE Transactions on Networking*, 16: 77–90, 2008.

[46] T. Spyropoulos, K. Psounis, and C. Raghavendra. Efficient routing in intermittently connected mobile networks: The single-copy case. In *ACM/IEEE Transactions on Networking*, 16: 63–76, 2008.

[47] T. Spyropoulos, K. Psounis, and C. S. Raghavendra. Spray and wait: An efficient routing scheme for intermittently connected mobile networks. In *SIGCOMM Workshop on Delay Tolerant Networking (WDTN)*, 2005.

[48] A. Vahdat and D. Becker. Epidemic Routing for Partially Connected *Ad Hoc* Networks. Technical Report CS-200006, Duke University, April 2000.

[49] J. Widmer and J.-Y. Le Boudec. Network Coding for Efficient Communication in Extreme Networks. *SIGCOMM Workshop on Delay Tolerant Networking (WDTN)*, 2005.

[50] J. Widmer, C. Fragouli, and J.-Y. Le Boudec. Energy-efficient broadcasting in wireless *ad hoc* networks. In *IEEE Workshop on Network Coding, Theory, and Applications (NETCOD)*, Riva del Garda, Italy, 2005.

[51] Wizzy Digital Courier. http://www.wizzy.org.za/.

[52] Y. Wu, P. A. Chou, and S.-Y. Kung. Minimum-Energy Multicast in Mobile *Ad hoc* Networks using Network Coding. In *IEEE Information Theory Workshop*, 2004.

[53] S.-Y. Kung Y. Wu, P. A. Chou. Information Exchange in Wireless Networks with Network Coding and Physical-Layer Broadcast. Technical Report MSR-TR-2004-78, Microsoft, August 2004.

[54] Q. Ye, L. Cheng, M. C. Chuah, and B. D. Davison. OS-multicast: On-demand Situation-aware Multicasting in Disruption Tolerant Networks. In *IEEE Vehicular Technology Conference (VTC)*, Montreal, Canada, 2006.

[55] X. Zhang, G. Neglia, J. Kurose, and D. Towsley. On the Benefits of Random Linear Coding for Unicast Applications in Disruption Tolerant Networks. In *NETCOD (IEEE Workshop on Network Coding, Theory, and Applications)*, 2006.

[56] X. Zhang, G. Neglia, J. Kurose, and D. Towsley. Performance modeling of epidemic routing. *Elsevier Computer Networks Journal*, 51/10: 2859–2891, 2007.

[57] X. Zhang, G. Neglia, J. Kurose, D. Towsley, and H. Wang. Random Linear Coding for Unicast Applications in Disruption Tolerant Networks. Technical Report 7277, INRIA, 2010. http://hal.archives-ouvertes.fr/inria-00494473/en/.

[58] F. Zhao, T. Kalker, M. Médard, and K. J. Han. Signatures for content distribution with network coding. In *IEEE International Symposium on Information Theory (ISIT)*, Nice, France, 2007.

[59] W. Zhao, M. Ammar, and E. Zegura. Multicasting in Delay Tolerant Networks: Semantic Models and Routing Algorithms. In *ACM Sigcomm workshop on Delay-Tolerant Networking (WDTN)*, Philadephia, PA, 2005.

INDEX

Printed in the United States
By Bookmasters